Anonymous

Inventors' and Mechanics' Scientific Manual and Patent Law

Complete

Anonymous

Inventors' and Mechanics' Scientific Manual and Patent Law Complete

ISBN/EAN: 9783337421328

Printed in Europe, USA, Canada, Australia, Japan

Cover: Foto ©berggeist007 / pixelio.de

More available books at **www.hansebooks.com**

INVENTORS' & MECHANICS'

Scientific Manual

AND

PATENT LAW COMPLETE.

PUBLISHED BY

The Western Union Patent Agency,

Peck & Miatt,

135 Clark Street, Chicago.

—

1871.

TO OUR

INVENTIVE FRIENDS

IN THE WEST.

We have compiled this little book to afford a convenient reference in patent and scientific matters for the benefit of the large and ever increasing class referred to,— a class which is doing more to advance the material prosperity of our country than any other.

We believe it will be acceptable to very many by answering for them many questions which would otherwise cause much trouble and expense to decide. It will also save us writing numberless long letters upon the subjects here treated of. Though we are ever ready to respond without charge in regard to any matter not fully discussed in these pages, still, if any one receives this book with a passage marked which fully answers his inquiry, we trust he will feel just as well satisfied with the answer as if it was written out.

The inventive talent of the West is being more and more largely employed; the harvest already reaped is rich in results, but the field has only just begun to reveal its extent, and it widens out before those who are looking eagerly to it with ever more inviting prospects. The truth is, most of us are just beginning to learn how to invent, and that there are better opportunities in this field to acquire rapid and honorable wealth and distinction than in almost any other. There can be no more honorable. work than that of invention. Every step made in the saving of time and in producing new and valuable effects means a certain amount of advance in the progress of civilization. When we have so far controlled material forces as to produce twice as much in the same time as is now produced, and also do it twice as well, we shall have gained an advantage which will be apparent in added comforts to almost every home in the land. Our fathers worked more hours and harder than we for smaller material results; our grandfathers received still less for their still harder toil. It is our desire to be of assistance in every way we can in helping on this general progress.

What has stimulated inventive genius especially in this country is the assurance which the laws give that every man shall have the sole right to whatever he originates for a considerable length of time, and this at little expense of money or time in obtaining the official guarantee of the government.

It is our especial part to see that the inventor has his invention secured to him in the most complete and advantageous manner.

Whatever facilities a long education in, and fondness for the work gives us, we offer to our many old and also our many prospective friends, promising them that aside from our regular charges in doing their work, we will gladly, and without charge, be of any assistance that we can in adding to, or informing them in any respect about, their devices.

RATIO of the hardness of metals: Iron, 1; platina, 2; copper, 3; silver, 4; gold, 5; tin, 6; lead, 7.

THE difference in time between high tide averages about 49 minutes each day.

ALL woods are from seven to twenty times stronger transversely than longitudinally. They become stronger both ways when dry.

NATURE AND EFFECT OF A PATENT.

WHEN one has perfected, in his own mind, some design of utility or beauty, the first thought is, I ought to be rewarded for this honest and skillful work, which is of such real worth. This result of the employment of my brain and time ought to be paid for by those who receive the benefit of it. Each who uses it to his or her advantage can well afford to pay me for providing it for them. It is for this exact want the governments of most enlightened countries have provided some form of Patent Right, which insures to the inventor the control and ownership of his invention for a limited time, which time is supposed to be sufficient for him in which to acquire the due reward of his exertions and skill.

In the United States the laws are more favorable to the inventor than in any other country. The fees for granting the patent are small, and there are no government annuities afterward to pay, while the protection to the inventor, so far as the law is concerned, is as ample as it can well be. The real value of the patent will, of course, depend upon the skill and knowledge with which the application has been prepared, but on any given patent the United States laws will always interpret in favor of the inventor where there is doubt upon any point. The provision is now so ample that any new form, device, or composition, can be secured to the originator under one of the following divisions, namely, a mechanical patent, a patent of design, a trademark, or a copyright.

Whoever neglects to avail himself of the provisions of the law — covering, as it does, every case — to retain to himself the benefit of the many new and bright ideas which suggest themselves to him, and allows the public generally, or perhaps some other inventor, to reap the benefits he might have enjoyed, has no one but himself to blame. And next, in lack of wisdom, to the man who neglects his opportunities, is he who goes before the Patent Office with a poorly or faultily prepared application. No money is better expended on an invention than that which is employed to have the application prepared in the best manner possible, so that technical delays and rejections may be avoided, and the patent issue in the best form to protect the inventor. This is where the skillful and experienced solicitor, who takes a real interest in the case, can be beneficially employed. We cannot do better than quote here the words of the late Commissioner of Patents, Hon. S. S. Fisher: "Inventors," he says, "are often poor, uneducated, and lacking in legal knowledge. They desire a cheap solicitor, and do not know how to choose a good one. They are pleased with the parchment and the seal, and are not themselves able to judge of the scope or value of the grant. Honest and skillful solicitors, with a thorough knowledge of the practice of the office and of patent law, and who are willing to advise their clients as to the exact value of the patents which they can obtain for them, may be of much service to inventors. There are many such, but those who care for nothing but to give them something called a patent, that they may secure their own fee, have, in too many instances, proved a curse. To get rid of their client and trouble they have sometimes been content to take less than he was entitled to. While in many cases they have, with much self laudation, presented him with the shadow when the substance was beyond his reach. Between such men and the office the strife is constant."

This puts the whole matter in a nutshell, and is good advice to those inventors who are inclined to think that every one who may call himself a patent solicitor is eligible to that business. It is a profession which requires a combination of different qualities from those demanded in ordinary business, and it is only those who concentrate all their powers upon it who are really competent to do the work.

We may say to those who have not yet employed us, that we have never lost a client to our knowledge, and that we have seldom failed to draw forth expressions of satisfaction, and often of thanks, for incidental assistance rendered them without charge in perfecting their inventions, or in securing patents where previous applications on the same subject had been rejected. Any inquiry may be addressed, with freedom, to PECK & MIATT, corner of Clark and Madison streets, Chicago, Ill.

THE assignee of every invention may have the patent issue to him directly, when he is the purchaser of the entire interest ; or the patent may issue in the joint names of the inventor and assignee, the inventor himself being one of the assignees.

THE application must be made by the actual inventor, if alive ; but if the inventor be dead, it may be made by his executors.

THE model must be neatly and substantially made of durable material, and not more than one foot in length or height.

A WORKING model is always desirable to enable the office to determine its precise operation. The name of the inventor and also of the assignee, if assigned, but be fixed upon it in a permanent manner.

No application can be examined, nor can the case be placed on file for examination, until the fee is paid.

WHAT is really embraced in the original invention, and so described or shown that it might have been embraced in the original patent, may be the subject of a re-issue.

THE mere fact of prior invention abroad will not prevent the granting of a patent, unless the invention had been described in some printed publication.

EVERY assignment should be recorded in the Patent Office within three months from its date.

ON THE CHARACTER OF INVENTIONS.

THE question is often asked us, what sort of an invention pays the best? Our reply always is, the devising of some simple thing that is universally needed, will bring the most rapid and proportionately the largest returns. The great inventions which require a long time to perfect and a long time to introduce, such as the telegraph, the steam engine, the vulcanizing of India rubber, etc., though they bring great fortunes and great names in the end, require a much greater expenditure of patience in giving proportionate results.

The small inventions pay the best, and nothing is too small to patent and to make money on, provided there is a sufficiently extensive use for it.

We are always ready to give our correspondents a careful opinion in regard to the value or novelty of any invention, as nearly as we can judge without taking too much time to look it up. When a special search or study is required $5 should be enclosed, that we may be able to afford sufficient time to the questions, to give them an exhaustive answer if the inquiries are such as to make that possible.

———

WHEN the patent is applied for here, after being obtained abroad, it will extend only fourteen years from the date of the foreign patent.

———

IF old materials and old principles are used in a state of combination to produce a new result, the inventor may obtain a valid patent for such result.

WHILE THE INVENTION IS BEING PERFECTED.

IT is oftentimes of great importance to have the invention protected while it is being tested or while the details of its working are being more fully finished. To give inventors every opportunity in this respect the law provides the caveat. The protection this affords is, to be sure, limited, but it is immediate. During its existence it prevents the issue of a patent for a similar device to any other person for at least three months after the caveator is officially notified that an application for a patent on his device or composition has been made. When the caveator files his application for a patent an interference may be declared by the Patent Office, in which case it is necessary to take proof as to who was the first inventor of what is equally claimed by both. The caveat is granted for one year, but may be indefinitely extended by paying the requisite fee of $10 at the end of each year.

The first cost of a caveat through the Western Union Patent Agency is $20, which includes the government fee and all charges. This is for an ordinary case. Where more than the usual amount of time is required in the preparation of the specification and drawings, the charge will be $25.

Caveats receive precedence over other business, and when specially requested we can have the papers ready to send to the inventor for his signature and affidavit by return mail. All that is required for us to properly prepare a caveat is a sufficient sketch or photograph to give the idea distinctly, and a general description with especial reference to the features it is desired to claim. Part of the fee, say $10, should also be sent at the same time as the sketch. No model is required.

PRELIMINARY EXAMINATIONS.

IT is frequently desirable to have a preliminary examination in the Patent Office, to show whether the invention is novel and patentable. As our charge for this service is only $5 (in advance), we usually advise it in all doubtful cases. If the search shows that nothing already exists to prevent the issue of a patent, the results of the examination will be of advantage to us in drawing up the application; and, if on the other hand, some previous invention is found to cover the same ground, the application fees, amounting to $45, will have been saved to the inventor. This examination will test the question of novelty, with the exception of caveats in the confidential archives of the Patent Office, a pending application before the office, a rejected application or the existence of a printed description of the invention. On receiving from our Washington office the written report of the search, we forward it, together with a sketch of any similar device which may exist, to our correspondent.

WHEN an application has been rejected, the inventor may manufacture his invention for two years, with the privilege of renewing his application within that time, if he so desires.

CORRESPONDENTS will please write their post-office address in plain, unmistakeable hand, at the bottom of their letter.

MODELS should be forwarded by express, and must invariably be prepaid.

THE PATENT.

The inventor, having so far perfected his idea that further delay seems inadvisable for the sake of greater perfection, before introducing it to the public and seeking a substantial return for his skill, looks eagerly to securing his patent. Upon the character of this will probably depend much of the success or failure of his invention when brought to the touchstone of public use. If his device is faultily or insufficiently or ambiguously embodied in the patent, it will probably be discovered by those interested, and litigation instead of profits may accrue to the owner. Too much care cannot, therefore, be taken to secure the most competent assistance in the preparation of the application. A re-issue may correct a previously defective patent, but that is paying for the patent twice over.

By the law of July 8, 1870, the *official fees are made the same to all, without distinction of nationality.* Patents are also granted on the same terms to women, minors, and executors or administrators.

The entire cost of an ordinary patent, when secured through us, is about $65. When an extra amount of time and labor are involved, a moderate charge, in accordance with the time, will necessarily be added, but we are rarely obliged to exceed the amount named. Of the $65, $17 is for first government fee, stamp and express, $20 for second government fee, leaving from $25 to $28 for our services in preparing the case, in presenting it at Washington, and in attending to its passage through the Patent Office.

We quote from the Patent Office rules as follows: "The personal attendance of the applicant at the Patent Office is unnecessary. The business can be done by correspondence or by attorney, and as the value of patents largely depends upon

the careful preparation of specifications and claims, the assistance of competent counsel will in most cases be of advantage
to the applicant."

When the model is ready—if the invention can be represented
by a model — send it by express to us, PECK & MIATT, corner
Clark and Madison streets, Chicago, Ill., inclosing also an
explanation of the working of the invention and of the particular features which it is desired to claim in the patent. Send
also, at the same time, in a separate package, by express or by
mail, in the form of a draft or postal order, the amount of the
first government fee, $17. On receipt of the case, as above, we
will proceed immediately to prepare the drawing, specification,
etc., and as soon as ready will forward the papers to the inventor for his signature and oath, accompanied also by our bill as
attorneys in the case.

It will thus be seen that the cost of making an application
for a patent is about $45. If the patent is allowed, a second
government fee of $20 is to be paid before it will issue.

The inventor should give his name in full.

THE inventor or inventors must sign the petition and claims,
the latter to be attested by two witnesses; also take the required
oath before some authorized person.

ACCORDING to a good authority — Wheatstone — an electric
spark travels through a copper wire at the rate of 280,000
miles in a second.

THE area of a circle is ascertained by multiplying the square
of its diameter by the decimal .7854, or by multiplying the
circumference by the radius (half diameter) and dividing the
product by 2.

SOUND travels in a still atmosphere at the rate of 1,120 feet
per second; so that the distance of a thunder cloud may be
known by multiplying the time in seconds between the flash
and the report by 1,120.

PATENTS OF DESIGN.

A PATENT of design will be granted to any person without distinction of nationality who, by his own genius, efforts or expense, has invented or produced any new pattern for the printing of woolen, silk, cotton or other fabrics; any original design for a bust, statue, etc.; for any impression, ornament, print, or picture, to be printed, painted, or otherwise worked into any article of manufacture; in fact, in any new and useful, or ornamental shape or configuration of any article of manufacture.

Patents for designs are granted for three and a half years, for seven years, and for fourteen years.

The whole cost of the patents, inclusive of our fees, are

For three and a half years $20
For seven years... 25
For fourteen years... 40

We require, to prepare the application for a design patent, the design itself, or if it can be shown distinctly by a photograph, two photographs (unmounted), together with the negative, and the name of the applicant. At the time of forwarding the design, which will generally be best done by express (prepaid), written directions should also be inclosed referring to those features which it is desired to cover in the patent, and also one-half the fee of $20, $25, or $40, according to the length of time for which the patent is desired. As soon as we can properly prepare the necessary papers, they will be sent to the applicant for his signature and oath; and the other half of the fee will then be due. When the application, with the proper signatures, is received, we forward it immediately to the Patent Office. An examination is there made, and if no conflicting design is found to exist, a patent is issued.

MODELS.

Every applicant for a patent is required by law to furnish a model in all cases where the invention can be wholly or in part illustrated in that way. Where the invention is an improvement on some known form or device, a working model of the whole will be unnecessary, but the working of the improvement should be distinctly shown. The dimensions of the model should not exceed twelve inches in any direction, and the material composing it should be of hard wood or metal, and the whole neatly executed. The name of the inventor and of the assignee (if assigned), and also the title of the invention, must be affixed upon it in a permanent manner. If two patents are to be taken on different parts of the same subject, a separate model for each will be required.

If the invention consists of a new composition, mixture or product, samples of the ingredients composing the compound should be furnished in sufficient quantities to make a specimen of the compound if desired, and these ingredients should be neatly done up and labeled. Almost all kinds of medical compounds, and useful mixtures of every description, are patentable. A very careful statement of the proportions of ingredients must accompany the same.

As a convenience to our patrons, we have made special arrangements for obtaining models to the best advantage in regard to both time and cost, and offer to procure them for those who cannot themselves obtain them so conveniently or cheaply. When a model is ordered, our friends will please to inclose a sketch or sufficiently distinct description to give us a clear idea of what is necessary, and, say $5, in addition to the $17, which would be required if the model had been sent.

Gutta-percha is readily dissolved by benzole, alcohol and bisulphuret of carbon.

TRADE-MARKS.

ANY person, firm or corporation, resident in the United States or any foreign country which, by treaty or convention, affords similar privileges to citizens of this country, and who are entitled to the exclusive use of any lawful trade-mark, may obtain protection for such trade-mark by complying with the official requirements.

It is necessary that affidavit be made by the applicant or some member of the firm or officer of the corporation that the party claiming protection for the trade-mark has an exclusive right to the use of the same, and that the description and *fac similes* presented for record are true copies of the trade-mark sought to be protected.

No proposed trade-mark will be received which cannot be a lawful trade-mark, or which is merely the name of a person or firm, unaccompanied by a mark sufficient to distinguish it from the same name when used by other persons, or which is identical with a trade-mark appropriate to the same class of merchandise, and belonging to a different owner, and already registered, or which so nearly resembles such last mentioned trade-mark as to be likely to deceive the public; but any lawful trade-mark already lawfully in use may be recorded.

The length of time which trade-marks remain in force is thirty years, and they may be extended for thirty years additional, except when claimed for and applied to articles not manufactured in this country, and in which it received protection under the laws of a foreign country for a shorter period, in which case it will cease to have force in this country at the same time that it becomes of no effect elsewhere.

The use of trade-marks is assignable. The assignment must be recorded within sixty days after its execution.

Certified copies of any trade-mark we can always obtain.

The cost of securing protection to a trade-mark is $35, of which $25 is for government fee, and $10 for our services.

Our correspondents who desire this protection will need to send us their full names, residence, and place of business, to state the class of merchandise and the particular description of goods in connection with which the trade-mark is to be used. Also, to describe the particular mode in which the trade-mark has been and is intended to be used, and to mention whether or not it has been in use, and if so, how long a time, and lastly, to inclose us six copies of the trade-mark. The fee of $35, in full for all expenses, should be inclosed at the same time with the order. On receiving the above directions, and fee, we will immediately prepare the petition, declaration, etc., and forward to the applicant for his signature and oath, and shortly after receiving the papers again will send the official certificate of protection.

The trade-mark furnishes the best and most ready means of protection to any article when a mechanical or design patent is not required or cannot be obtained.

———

IF a piece of timber that has been for a long time exposed to water be brought into the air and dried, it will become brittle and useless.

———

THE area of an ellipsis is found by multiplying the long diameter by the short diameter and dividing the result by the decimal .7854.

———

THE application must be made by the actual inventor, if alive, even if the patent is to issue or re-issue to an assignee; but where the inventor is dead, the application and oath may be made by the executor or administrator.

———

JOINT inventors are entitled to a joint patent; neither can one claim separately; but independent inventors of separate improvements in the same machine cannot obtain a joint patent for their separate inventions; nor does the fact that one man furnishes the capital and the other makes the invention, entitle them to take out a joint patent.

PATENTS TO WOMEN AND MINORS.

As before stated, women and minors may obtain mechanical and design patents, trade-marks and copyrights.

The law makes no distinction as to age or sex, and we are pleased to see that both women and young America are more and more improving their opportunities in producing good and valuable inventions. For the ready talent of both there is ample field.

We have often thought how much of value might have been created already if the ready wit and subtle intellect of women had been applied to the subject of invention. It is also eminently desirable that those who hope to attain to success of the best kind in inventing should habituate their minds to the necessary study, observation and habits of thought while young. The inventive faculties are largely educative. There are thousands who are now dragging out a life of toil and poverty who, if they had applied their minds to studying and thinking of what was about them, would some time have hit upon an improvement or invention which would have changed their poverty to competency and perhaps distinction. This leads us to say that there are many persons who think they have no inventive talent. In most cases it is as if a man who did know the alphabet should say that he had no talent for rhetoric. As in other things, some people undoubtedly have superior gifts in this direction as compared with others, but we believe there is no set of faculties susceptible of more enlargement than those involving invention.

RULE 23 of the Patent Office says: "Applicants are advised to employ competent artists to make the drawings, which will be returned if not executed in strict conformity with these rules, or if injured by folding."

COPYRIGHTS.

ANY citizen of, or resident in the United States, who is the author, designer or proprietor of any book, map, chart, engraving, cut, print, photograph or negative thereof, or of a painting, drawing, chromo, statue or statuary, a dramatic or musical composition, or designs or models, intended to be perfected as works of the fine arts, may obtain a copyright thereon. But a copyright cannot be had unless the title or description is recorded in the library of congress before the publication of the work. Non-resident foreigners cannot obtain copyrights, but foreigners who reside in the United States may obtain them.

Copyrights are granted for twenty-eight years, with a renewal term of fourteen years additional.

Copyrights can be assigned. The assignment must be recorded by the librarian of congress.

The infringement of copyrights is the subject of heavy fines and penalties.

The first step for one who desires a copyright is to send us a printed title of the book, paper, photograph or article, together with the whole cost in the case, namely, $6. We will then forward the case promptly to Washington, and shortly thereafter send the applicant the official certificate of copyright

Applicant must then send us three copies of the first issue of the copyrighted article as soon as produced, under a government penalty of $25 fine for neglect so to do. This comprises all the trouble and expense in securing a copyright. Address us for any further information.

AMENDING AND APPEALING.

WHEN an application for a patent is passed upon adversely by the examiner at the Patent Office, we draw up new arguments and present the case anew, in what we believe to be the best form to obtain his allowance, and at the same time protect the rights of our client.

This we do in any case in hand without additional charge. Should the examiner reject the case on its second presentation, we report the fact promptly to our client, accompanied with our estimate of the probabilities of obtaining a reversal of the examiner's decision by an appeal to the board of examiners-in-hief. For this appeal to the board the government fee is $10. Our own charge is very moderate for the amount of labor involved. If desired, we will make our fee dependent on success.

Should the application not be allowed by the board, a second appeal may be taken to the commissioner of patents; the government fee being $20. Our charge, as before, will be small, or contingent on success.

Appeal may also be taken, except in interference cases from the commissioner of patents, to the supreme court of the District of Columbia. Costs payable by applicant.

It is always trying to the patience of one in haste for his patent to be obliged to appeal, and, accordingly, we not only do everything in our power, professionally, but always make our charge for the necessary labor as light as possible.

IF the inventor, at the time of making his application, believes himself to be the first inventor or discoverer, a patent will not be refused on account of the invention or discovery, or any part thereof, having been before known or used in any foreign country: it not appearing that the same or any substantial part thereof had before been patented or described in any printed publication.

LAPSED CASES.

THIS refers to such applications for patents as have been "allowed," but the final government fee not paid within the six months allowed by law. All such cases can be renewed at any time within two years from the date of the original allowance, by filing a new application and paying the same government fees as on the original application. Our own fee in such cases, where the previous application was well prepared, is but trifling.

―――――

REJECTED CASES.

THE number of these is already large, and as the volume of patent business increases, we have more and more brought to us. Without exactly making a specialty of this branch, we still confess to taking great pleasure in securing patents for those who had almost lost hope of obtaining them. When the invention is in itself patentable, rejections occur from two causes, namely, improperly prepared papers, and errors of the Patent Office. From one of these causes rejections often occur when the real merit in the application, if skillfully presented, would be rewarded by a patent. We have rare facilities, through our office in Washington, for presenting rejected cases favorably to the Patent Office, and are always glad to undertake such cases, either without charge, except in case of success, or on the ordinary fee for the time and labor employed. We particularly invite the correspondence of all who have such cases, from the confident feeling we have that if it is possible for anything to be done for them, that we can do it.

RE-ISSUES.

RE-ISSUE is granted to the original patentee, his legal representatives, or the assignees of the entire interest, when, by reason of a defective or insufficient specification, the original patent is inoperative or invalid, provided the error has arisen from inadvertence, accident or mistake, and without any fraudulent or deceptive intention.

The general rule is, that whatever is really embraced in the original invention, and so described or shown that it might have been embraced in the original patent, may be the subject of a re-issue; but no new matter can be introduced into the specification, nor in case of a machine patent can the model or drawings be amended except each by the other; but, when there is neither model nor drawing, amendments may be made upon proof satisfactory to the commissioner that such new matter or amendment was a part of the original invention, and was omitted from the specification by inadvertence, accident, or mistake, as aforesaid.

Re-issued patents expire at the end of the term for which the original patent was granted. For this reason applications for re-issue will be acted upon as soon as filed.

A patentee, in re-issuing, may, at his option, have a separate patent for each distinct and separate part of the invention comprehended in his original patent, by paying the required fee in each case, and complying with the other requirements of the law, as in original applications. Each division of a re-issue constitutes the subject of a separate specification descriptive of the part or parts of the invention claimed in such division, and the drawing may represent only such part or parts. All the divisions of a re-issue will issue simultaneously. If there be controversy as to the one, the other will be withheld from issue until the controversy is ended.

Very many patents are being re-issued in order to cover more perfectly the inventions to which they relate.

We offer our advice and ask the correspondence of those whose patents are not now in a form to suit them.

The ordinary cost of a re-issue, when obtained by us, is about $60, including $30 for government fees. We shall be pleased to give any further information.

EXTENSIONS.

POWER is vested in the commissioner to extend any patent granted prior to March 2, 1861, for seven years from the expiration of the original term; but no patent granted since March 2, 1861, can be extended. When a patent has been re-issued in two or more divisions, separate applications must be made for the extension of each division.

The questions which arise on each application for an extension are :

First. Was the invention *new* and *useful* when patented?

Second. Is it *valuable* and *important to the public*, and to what extent?

Third. Has the inventor been *reasonably remunerated* for the time, ingenuity, and expense bestowed upon it, and the introduction of it into use? If not, has his failure to be so remunerated arisen from neglect or fault on his part?

Fourth. What will be the effect of the proposed extension upon the public interests?

The applicant for an extension must file his petition and pay in the requisite fee not more than six months nor less than ninety days prior to the expiration of his patent. There is no power in the commissioner to renew a patent after it has once expired.

INTERFERENCES.

AN "interference" is an interlocutory proceeding for the purpose of determining which of two or more persons, each or either of whom claims to be the first inventor of a given device or combination, really made the invention first.

An interference will be declared in the following cases :

First. When the parties have pending applications before the office at the same time, both or all the parties claiming to be the inventor of the same thing.

Second. When an applicant, having been rejected upon the prior unexpired patent or the prior application of another, claims to have made the invention before the prior applicant or patentee.

Third. When an invention is claimed in a renewed application which is shown or claimed in an application filed, or unexpired patent granted prior to the filing of such renewed application.

Fourth. When an applicant for a re-issue embraces in his amended specification any new or additional description of his invention, or enlarges his claim, or makes a new one, and thereby includes therein anything which has been claimed in any patent granted subsequent to the date of his original application, as the invention of another person, an interference will be declared between the application and any such unexpired patent or pending application. If the re-issue application claims only what was granted in the original patent it may be put into interference with any pending application in which the same thing is shown, provided the latter applicant claims to be the prior inventor, and is not barred a patent by public use or abandonment.

Fifth. When an application is found to conflict with a caveat the caveator is allowed a period of three months within which to present an application, when an interference may be declared.

Sixth. The office reserves to itself the right, when two applications are pending at the same time, in one of which a device may be described which is claimed in the other, to declare an interference to determine with whom is priority of invention without reference to the order in which such applications may have been filed.

DISCLAIMERS.

WHENEVER, by inadvertence, accident, or mistake, the claim of invention in any patent is too broad, embracing more than that of which the patentee was the original or first inventor, some material or substantial part of the thing patented being truly and justly his own, the patentee, his heirs or assigns, whether of a whole or of a sectional interest, may make disclaimer of such parts of the thing patented as the disclaimant shall not choose to claim or to hold by virtue of the patent or assignment, stating therein the extent of his interest in such patent; which disclaimer must be in writing, attested by one or more witnesses, and recorded in the Patent Office.

ASSIGNMENTS.

A PATENT may be assigned, either as to the whole interest or any undivided part thereof, by any instrument of writing.

A patent will, upon request, issue directly to the assignee or assignees of the entire interest in any invention, or to the inventor and the assignee jointly, when an undivided part only of the entire interest has been conveyed.

The patentee may grant and convey an exclusive right under his patent to the whole or any specified portion of the United States, by an instrument in writing.

Every assignment or grant of an exclusive territorial right must be recorded in the Patent Office within three months from the execution thereof; otherwise it will be void as against any subsequent purchaser or mortgagee for a valuable consideration without notice; but, if recorded after that time, it will protect the assignee or grantee against any such subsequent purchaser, whose assignment or grant is not then on record.

The patentee may convey separate rights under his patent to make or to use or to sell his invention, or he may convey territorial or shop rights which are not exclusive. Such conveyances are mere licenses, and need not be recorded.

HEARINGS.

ALL cases pending before the commissioner will stand for argument at one o'clock on the day of hearing. If either party in a contested case, or the appellant in an *ex parte* case, appear at the time, he will be heard; but in contested cases no motions will be heard in the absence of the other party, except upon default after due notice ; nor will a case be taken up for oral argument after the day of hearing, except by consent of both parties. If the engagements of the tribunal before whom the case is pending are such as to prevent it from being taken up on the day of hearing, a new assignment will be made, or the case will be continued from day to day until heard. Unless otherwise ordered before the hearing begins, oral arguments will be limited to one hour for each counsel.

LAYING OUT MACHINERY AND ESTIMATING COST.

WE make it a part of our business to act for our correspondents as mechanical engineers in arranging the details, size, etc., of machinery required to do a given work, and in making careful estimates of the cost of same.

Many inventors only build models of their improvement previous to obtaining the patent, and when they come to require the actual working devices, they need competent assistance in drawing the plans for the proportionate size and strength of the different parts, as well as advice as to the best manner of construction   This service, when entrusted to us, will always be rendered with fidelity, and the charge will be, in all cases, moderate, dependent in part upon the amount of assistance rendered, but having reference, sometimes, to the time employed.

It is also our practice to attend to purchasing and shipping anything mechanical our correspondents may desire, and for this we make as small a charge as is consistent with the trouble involved. If our clients only send distinct directions, and what money is required in any particular case, their commissions will be executed in the best manner.

Those who require plans for machinery or any particular device, will be careful to give us a full explanation of the character and amount of the work to be performed, or purpose which is to be served, and to enclose us a model, sketch or other means of determining the exact character of the article required.

It is often desirable to have an invention carefully examined by a competent mechanical engineer previous to making an application for a patent upon it. Many inventors have found their machines worked well in a model or on a small scale, but when the attempt has been made to build large working machines on the same plan, they have been found defective, and surprise as well as disappointment has ensued. There are many causes which go to bring about this result, which only those who are well versed in the theory and practice of mechanics, and familiar with the action of physical laws and the strength of materials, can understand, or are apt to think of. The disproportion of parts, unequal or misplaced strain, working out of line, the effects of expansion and contraction are a few of the many common causes of unsatisfactory working in machinery. When employed to secure the patent or patents on a machine, we make no charge for general advice; and when the inventor desires us to furnish him with working plans for the same we take into account, in making our charge, the previous business furnished us.

THE PROGRESS OF INVENTION.

To illustrate how interest is increasing in the subject of invention, we make a short extract from the report of the commissioner of Patents for 1868. He says, "During the year ending December 31, 1868, there have been filed in the Patent Office 3,705 caveats and 20,445 applications for patents; 12,959 patents have been issued, 419 have been re-issued, and 140 extended.

Compared with other years the business of the office has been greater than that of any preceding period. The number of patents issued has been more than double the number of 1865, and more than three and a half times that of 1858.

Since the Patent Office was first established its business has had a rapid growth in amount and in importance. In 1836 eight or ten persons were enough to transact all its business. Now between three and four hundred are required.

This increase has arisen in part from the growth of the country, but more from the stimulus that our patent laws have given to invention. The reward which they have held out for successful improvements have increased in value with the progress of the country, and with the more proper appreciation and greater security of patented property. A really successful invention now brings to its author a competency for life; and, as a consequence, the efforts of almost every class in the community are directed in the line of useful improvements.

In all those improvements of life to which patent laws relate, our own age has witnessed more advance than all the preceding ages of the world taken together. One improvement seems to have begotten another. New fields for exploration have been constantly opening. And so far from reaching any limit of invention, we seem but on the way to other advances and improvements beyond our present comprehension.

EXPLAINING INVENTIONS.

INVENTORS sometimes find themselves greatly perplexed while endeavoring to illustrate and explain their invention to others, very frequently those whom they wish to interest in the subject, for assistance in obtaining a patent, or working it after it is allowed, simply because the drawings or rough sketches they exhibit for examination are so vague and incorrect that none but the maker can understand the jumbled combination of lines.

In disposing of, or describing an invention for any purpose, a neat, artistic and correct drawing illustrating the device in all its mechanical relations and proportions is invaluable, not only because it saves the explainer's time and temper, but being easily understood at a glance, and the artistic execution and symmetry of proportion pleasing to the eye, it produces a favorable impression in the observer's mind at once, and the natural prejudice every man feels against a thing he does not, or cannot understand, is almost overcome at the outset.

On the other hand, if an inventor unrolls a poorly executed, incorrectly proportioned sketch of the invention, no matter how valuable or simple the device may be itself, the observer is very apt to judge it as being complicated and impracticable, and shrinks from a further acquaintance with it, simply because he does not understand it.

Again, inventors might frequently save themselves considerable time and labor if they had a plan made of their ideas as developed on a working scale, so that every part could be properly proportioned and arranged in a practical manner, and all errors of construction avoided. A great many ideas that seem practical on a small scale, or to the inventor, when thoroughly examined by a mechanical engineer, are found to be incorrect in some detail or other.

There are also numerous other instances in which the assistance of competent mechanical draughtsmen and advisors would be of untold benefit, both in saving the time and money of the inventor, by preparing for him plans of his invention on a correct working scale, and pointing out to him any defects or impracticabilities in his device, that those who have had long experience, and a special education in engineering and mechanics are alone able to perceive. Having had the advantage of long daily experience and education in this direction, and having an efficient and careful department of draughtsmen in connection with our business, we feel able to offer our clients, and inventors and others interested, our services in preparing drawings, working-plans or scales, colored drawings in every branch of mechanical draughting, and at rates very moderate for the work performed; while we will freely point out any defect or render any assistance to the inventor in our power to make his invention effective and practical, without charge.

TRANSMISSION OF HEAT BY VARIOUS SUBSTANCES.—Though the powers of bodies capable of transmitting heat and light are not in the same relative proportions, yet all which transmit heat are more or less transparent, as will be seen by the following, as given by Melloni:

Air	100	Rape Seed Oil	2
Rock Salt, transparent	92	Tourmaline, green	7
Flint Glass	67	Sulphuric Ether	21
Bisulphuret of Carbon	63	Gypsum	20
Calcareous Spar, transparent	62	Sulphuric Acid	17
Rock Crystal	62	Nitric Acid	15
Topaz, brown	57	Alcohol	15
Crown Glass	49	Alum in Crystals	12
Oil of Turpentine	31	Water	11

REFLECTING POWERS OF METALS, ETC.—The best reflectors of heat are the metals. The laws of the reflection of heat are the same as those of light — the angle of reflection being opposite and equal to that of incidence. We give the following by Leslie:

Brass	100	Steel	70
Silver	90	Lead	60
Tin Foil	85	Glass	10
Block Tin	85	Glass, waxed or oiled	5

CONDUCTING POWERS OF METALS.—Metals stand the highest in conducting power, and wood among the lowest, the softer kinds of wood being the lowest. The following is according to Despritz:

Gold	1,000	Tin	304
Silver	973	Lead	180
Copper	898	Marble	24
Platinum	381	Porcelain	12
Iron	374	Tile	11
Zinc	363		

RADIATING POWER OF METALS, ETC.—Polished surfaces radiate heat less than those which are rough, and those substances which are the best conductors of heat are generally the poorest radiators, as the following list, on the authority of Leslie, will show:

Lamp-Black 100	Rough Lead.............................. 46
Water...100	Mercury................................... 20
Writing Paper............................. 98	Polished Lead.......................... 19
Glass.. 90	Polished Iron........................... 15
Tissue Paper................................ 88	Tin, Silver, Copper and Gold.......... 12
Ice.. 35	

VELOCITY AND POWER OF WIND.—To ascertain the force of the wind acting perpendicularly upon the plane surface, multiplying the surface in feet by the square of the velocity in feet, and the product by .002288. This will give the force in pounds, avoirdupois:

Miles Per Hour	Feet Per Second.	Force in Pounds, Avoirdupois, Per Square Foot.	
1	1.47	.005	Scarcely perceptible.
2	2.93	.020	
3	4.40	.044	Just perceptible.
4	5.87	.079	
5	7.33	.123	Gentle wind.
10	14.67	.492	
15	22 00	1.107	Brisk wind.
20	29.34	1.968	
25	36.67	3.075	
30	44.01	4.429	High wind.
35	51.34	6 027	
40	58.68	7.873	Very high.
45	96.01	9.963	
50	73 35	12 300	A storm or gale.
60	88.02	17.715	A great storm.
80	117.36	31.490	A hurricane or tempest.
100	146.70	49 200	{ A violent hurricane sufficient to uproot trees and level houses.

UNDER ordinary circumstances sound travels through air at the rate of 372 yards per second; through water about four times as fast as through air; through acacia wood at the rate of 5,142 yards per second; through deal, 3,630 yards; through poplar, 4,670 yards; through oak, 4,200 yards; through ash, 5.090 yards.

Some metals transmit sound still more rapidly. In an iron wire, for example, the rate is 5,363 yards per second, and in steel. 5.436 yards, while through brass the velocity is only 3,888 yards.

THE celebrated axle grease invented by Mr. Booth is made as follows: Dissolve ½ pound common soda in 1 gallon water, add 3 pounds tallow and 8 pounds palm oil (or 10 pounds palm oil only). Heat them together to 212° Fahrenheit, mix and keep constantly stirred till the composition is cooled down to 70°.

THE utmost velocity that can be given to a cannon ball is 2.000 feet per second. It order to increase the velocity from 1,650 feet per second — the ordinary rate — to 2,000, one-half more powder is required.

THE greatest natural temperature ever authentically recorded was at Bagdad in 1819, when the mercury in Fahrenheit's thermometer rose to 120° in the shade. Buckhardt, in Egypt, and Humboldt, in South America, observed it at 117° Fahrenheit in the shade.

About 70° below zero, in Fahrenheit's thermometer, is the lowest atmospheric temperature reached by Arctic navigators. The greatest artificial cold ever produced was 220° Fahrenheit, below zero. At this temperature neither alcohol or ether were frozen.

The temperature of the space above the earth's atmosphere has been estimated at 58° below zero, Fahrenheit.

AIR is highly compressible and perfectly elastic. By these two qualities air and all other gaseous substances are particularly distinguished from liquids, which resist compression and possess but a small degree of elasticity. The density and elasticity of air are directly as the force of compression.

The volume of space which air occupies — and the rule is the same for almost all gases — is inversely as the pressure upon it. If the compressing force be doubled, the compressed air will occupy one-half the former space.

Like liquids, all aeriform or gaseous substances transmit pressure equally in every direction.

The amount of pressure exerted by the atmosphere at the level of the ocean is 15 pounds per square inch of surface.

To petrify wood, etc., make a mixture of equal parts of chalk, white vinegar, gem-salt, rock alum and powdered pebbles. After the ebullition which ensues has subsided, immerse in the liquor whatever porous matter it is desired to petrify, and allow it to soak about six days, when it will be found to have turned into a petrifaction.

DUCTILITY AND MALLEABILITY OF METALS. — Metals which draw out into the finest wire are not those which afford the thinnest leaves under the hammer, or in passing through the rolling-press. Iron is a good illustration. The most ductile cannot be wire-drawn to any considerable extent without being annealed from time to time during the process of extension. This enables the particles to slide along side of each other so as to loosen their lateral cohesion.

Seventeen of the metals which retain their metallic form in the air are ductile, and sixteen are brittle.

WATER boils in a vacuum at 98° Fahrenheit, if the vacuum is nearly perfect.

THE bulk of tallow is 50 pounds in a cubic foot; of oakum, 12 pounds in a cubic foot; of oil, 6.23 gallons in a cubic foot; and coal, 45 cubic feet in a ton.

A GOOD general rule for calculating the strength of hempen cables is to multiply the square of the circumference in inches by 120, and the product will be the weight in pounds the cable will safely support. For a rope, multiply the square of the circumference in inches by 200, and it gives the weight the rope will safely bear. When the rope, cable or hawser is made of Manilla hemp, the weight of a single foot is approximately ascertained by multiplying the square of the circumference in inches by the decimal .03.

COMPARATIVE POWER OF MAN OR HORSE AS APPLIED TO MACHINERY.—A man is estimated to exert a force, in lifting or carrying, of 6,000 pounds, one foot per minute; in turning the winch of a crane, 6,300; in pumping, 3,814; in ringing, 8,570;

in rowing, 9,010. The power of a horse is equal to 33,000 pounds, raised one foot per minute, or 150 pounds at the rate of 220 feet per minute.

GOOD gunpowder is composed very nearly of 1 equivalent of nitre, 3 of carbon, and 1 of sulphur. Much of the explosive energy of gunpowder depends on its granulation; a fine dust of the same composition as powerful powder burns rapidly, but without explosion.

THE height of clouds is exceedingly variable, and their mean elevation is not the same in different countries. The stratus cloud often descends to the earth's surface. In pleasant weather the lower limit of cumulas clouds varies from 3,000 to 5,000 feet elevation, and the upper limit from 5,000 to 12,000 feet. Cirrus clouds are never seen below the summit of Mount Blanc, which has an elevation of more than 15,700 feet.

DEW is produced most copiously in tropical countries, because there is the greatest difference between the temperature of the day and that of the night. Upon the small islands of the Pacific dew rarely forms, the air over the vast ocean preserving a nearly uniform temperature day and night.

IT is highly probable that the cirrus cloud at great elevations does not consist of vesicles of mist, but of flakes of snow.— *Wells.*

THE remark is sometimes made that "it is economy to burn green wood." This idea is an error. Green wood burns less rapidly, but to produce a given amount of heat a greater amount of fuel must be consumed. The moisture must be evaporated before it will burn, and an additional amount of fuel will be required to vaporize the sap.

AT the equator in Brazil the average annual temperature is 84°, Fahrenheit's thermometer : at Calcutta, lat. 22°, 35′ N., it is 78° F. : at Savanna, lat. 32°, 5′ N., it is 65°; at London, lat. 51°, 31′ N., it is 50° ; at Melville Island, lat. 74°, 47′ N., it is 1° below zero.

AIR at 32° Fahreneit, can absorb the 160th part of its own weight, and, for every 27 additional degrees, its capacity for absorbing moisture is double that at 31°.

ON a clear night in summer, when dew is depositing, the mercury in a thermometer laid in the grass will sink nearly 20° below that in one suspended in the air.

LIQUIDS expand under the influence of heat to a greater degree and more unequally than solids. A column of water contained in a cylindrical vessel will expand 1-23 in length when heated from the freezing to the boiling point. A column of iron, with the same increase in temperature, will expand only 1-84 its length. Spirits of wine, with an increase from 32° to 212°, gains 1-9 in bulk; oil, 1-12.

VAPORS rise into any space filled with air in the same manner as if air was not present, the two fluids seeming to be independent of each other. A vessel filled with air will receive as much vapor of water as one from which the air has been exhausted.

OF the two gasses which compose air, oxygen forms one-fifth and nitrogen four-fifths.

THE weight of wood varies greatly. A cord of dry hickory weighs 4,400 pounds; a cord of soft maple, 2,600 pounds.

IT is an exceedingly curious fact that one law applies to all physical influences which spread from a centre, such as gravitation, heat, light, sound, electrical forces, and all central forces, when not weakened by any resisting or opposing force, namely, that the intensity varies inversely as the square of the distance; that is, at twice the distance from the source the influence or effect is only one-fourth as great; at three times the distance only one-ninth, and so on.

ALL gasses and aeriform substances expand 1-490 for every degree of heat they receive above 32° Fahrenheit, and contract in the same proportion for every degree below that point.

EVERY 550 feet above the sea level decreases tne boiling point of water 1° Fahrenheit. In the city of Quito, in South America, water boils at 194° Fahrenheit, instead of 212° F.; its height above the sea level is, therefore, 9,541 feet.

IN the temperate zone the average fall of rain is 35 inches in a year, and in the tropics, 95 inches.

GREASE spots may be extracted from paper by sprinkling with powdered pipe-clay and then applying a hot iron; afterwards removing the powder with a piece of India-rubber.

THE celebrated cement used by American jewellers to fasten precious stones on to the plain surface of gold and silver is made as follows: Isinglass (Russian) soaked in water till softened a little (but none of the water must be used), to the amount of five or six pieces, and finally dissolve in two ounces of French brandy. The solution should be thick. Dissolve in this ten grains of very pale gum ammoniac (in tears) by rubbing them together. Then add six large tears of gum mastic, dissolved in the least possible quantity of rectified spirits. Mix well with sufficient heat.

WATER rises in a suction pump in proportion as the pressure of the atmosphere, which is fifteen pounds to the inch, is removed. For this reason water can never rise in this kind of pump to a greater height than thirty-four feet.

THE temperature increases, as we descend into the earth, at the rate of one degree for every fifty feet.

ARCTIC explorers, while breathing air that freezes mercury, still have in them the natural warmth of 98° Fahrenheit, above zero; and the inhabitants of India and Arabia, where the mercury sometimes stands at 115° Fahrenheit, in the shade, have their blood at no higher temperature. Of all animals man alone is capable of living in all climates, and of changing his place of abode to all portions of the earth.

LIQUID and gaseous bodies are almost absolute non-conductors of heat.

GREEN sealing-wax can be made by powdering and mixing slowly, under heat, two parts of shellac, one of yellow resin, and one of verdigris.

Black, by mixing, in the same way, three parts yellow resin, two shellac, and two of ivory-black.

Gold-colored, by mixing slowly one pound bleached shellac, four ounces Venice turpentine and gold-collored talc, as required.

Red, with two parts shellac, one of resin, and one of smalts.

Marbled, by mingling the above colors when they begin to cool.

A GOOD ink for writing on steel, tin-plate or sheet-zinc is made by mixing one ounce of powdered sulphate of copper and half an ounce of powdered sal ammoniac, with two ounces of diluted acetic acid, and adding lamp-black or vermillion.

A BODY descending from a height will fall sixteen feet in the first second, three times that distance in the second, and so on, increasing as the odd numbers 1, 3, 5, 7, 9, etc.

The space passed over by a falling body is as the square of the time: in twice the time it will fall four times the space, etc.

When the time occupied in falling is known, the height from which the body falls may therefore be known by multiplying the square of the number of seconds of time consumed in the descent by sixteen, — the distance a body will fall in one second of time.

OF all transparent bodies the diamond possesses the greatest refraction or light-bending power, although it is exceeded by a few deeply-colored almost opaque minerals. It is mostly from this property that the diamond derives its brilliancy.

COPYING ink is easily made by adding one ounce of moist sugar to a pint of common ink.

COUNT RUMFORD, by his experiments, made over sixty years

ago, proved that if powder was placed in a close cavity. and the cavity two-thirds filled, its dimensions being at the same time restricted, the force of explosion would exceed 150,000 pounds upon the square inch.

THE power exerted in performing a certain work is equal to the weight of the body moved in pounds multiplied by the vertical space through which it is moved.

THE strongest of all metals for resisting tension or a direct pull is tempered steel.

A GOOD GREEN INK is made by mixing 1 part of cream of tartar, 2 parts verdigris, and 8 parts water, boiling until the proper color appears.

THE pressure exerted by a column of liquid is equal to, and measured by, the height of the column, and not by its bulk or quantity.

SOUND decreases from the point where it originates, according to the law by which the attraction of gravitation varies, viz. : inversely as the square of the distance, Thus, at double the distance it is only one-half as strong; at three times the dis tance, one-ninth.

A SHORT pipe will discharge one-half more water in the same time than a simple orifice of the same dimensions.

COMMON BLACK INK may be given an intense jetty color by adding a little impure carbonate of potassa.

A SOLID immersed in a liquid will be pressed upward with a force equal to the weight of the liquid it displaces.

AN inch tube 200 feet in length placed horizontally will discharge only one-fourth as much water as a tube of like size one inch in length.

IN constructing a room for public speaking the ceiling should not be more elevated than 30 or 35 feet.

WATER is about 840 times heavier than air, taken bulk for bulk. The weight of the atmosphere enveloping the earth has been estimated to be equal to the weight of a globe of lead sixty miles in diameter.

To make artificial coral, melt together 4 parts of resin and 1 part of vermillion.

THE range of projectiles is greatest, in case of high velocity, when the range of elevation is 30°; but for slow motions, the horizontal distance attained is greatest when the angle of elevation is 45°.

THE following shows the average weights sustained by wires of different metals, each having a diameter of about one-twelfth of an inch: Lead, 27 lbs.; zinc, 109 lbs.; silver, 187, copper, 303; tin, 34; gold, 150; platinum, 274; iron, 549, steel, 1,010. Cords of different materials, but of the same diameter, possess the following relative strength: Common flax, 1,175 lbs.; New Zealand flax, 2,380 lbs.; hemp, 1,633 lbs.; silk, 3,400 lbs.

In a large room, nearly square, the best place to speak from is one corner with the voice directed diagonally to the opposite corner. It is better generally to speak from pretty near a wall or pillar.

WAVES have been observed to rise to a height of about 43 feet above the hollow occupied by a ship; the distance between the crests of two large waves being 559 feet, and the time occupied by a wave in passing this distance 17 seconds.

THE height to which the atmosphere extends above the surface of the earth is believed upon good grounds to be about 50 miles.

THE heat of the sun penetrates into the earth varying distances, according to the nature of the surface, from 50 to 100 feet.

WATER is composed of 8 parts of oxygen and 1 of hydrogen.

These gases, when produced separately and then united, make the most powerful artificial heat known. An economical process of decomposing water into its constituent gasses would be one of the most valuable inventions possible. This can already be done on a small scale by electricity—one pole of the battery sending off bubbles of hydrogen, and the other oxygen, when the two poles are brought near together.

THERE is no motion in the universe without a corresponding and opposite action of equal amount. Action and reaction are always equal. A man in rowing impels the water astern with the same force that he drives the boat forward.

THE gold lining of Chinese cabinets may be closely imitated, according to Dr. Wiederhold, of Cassel, by the following formulæ: First of all, two parts of copal and one of shellac are to be melted together to form a perfectly fluid mixture; then two parts of good boiled oil, made hot, are to be added; the vessel is then to be removed from the fire and ten parts of oil of turpentine are to be gradually added. To give color, the addition is made of solution in turpentine of gum-gutta, for yellow, and dragons' blood for red. These are to be mixed in sufficient quantity to give the shade desired.

A PHYSICIAN in Florence cures somnambulism by winding once or twice around the patient's leg, on retiring, a thin, flex-ible copper wire, long enough to reach the floor.

THE moon's path in the heavens is not a circle, but a continuous spiral, that does not return into itself until after a very long time. In consequence of its spiral movements along the face of the heavens the moon hides in succession every point in a belt of sky that stretches to five degrees, eight minutes and forty-eight seconds on either side of the ecliptic, and it takes eighteen and a-half years to do so

A NEW difficulty has occurred in the practical working of the Suez Canal. The heat is so great that the stokers cannot live through it. Many have died in the passage, and all tell the same story of terrible suffering. Climatological maps show that although neither the Isthmus of Suez nor the Red Sea is equatorial, the "district of greatest heat" throughout the whole globe is a small space which crosses the Red Sea, Arabia and the Persian Gulf.

THE momentum or force which a moving body exerts is determined by multiplying its mass or weight by its velocity. Thus, a body moving at the rate of 100 feet per second and weighing 25 lbs., will strike with a force of 2,500 lbs.

THE various rays composing sun light are not equally luminous, that is they do not appear to the eye as equally brilliant The color best seen by the human eye is yellow. The intensity of the different colored rays of light expressed numerically is: red, 94; orange, 640; yellow, 1,000; green, 480; blue, 170; indigo, 31; violet, 6.

The proportionate number of soldiers, wearing different colors, who are killed in battle, appears from numerous observations to be as follows: Red, 12; dark green. 7: brown, 6; blueish gray, 5. Red is, therefore, the most fatal color, and a light gray the least so.

IN a vacuum all bodies fall with equal velocities. A bullet and a feather dropped at the same time will reach the bottom of a vessel exhausted of air at the same moment.

THE compass is claimed to have been discovered by the Chinese. It was known, however, in Europe, and used in the Mediterranean in the thirteenth century. A compass at that time consisted of a piece of loadstone fixed to a cork, and floating on the surface of water.

AS the most exact calculation of the speed of light, founded on perfectly accurate data, gives 192,500 miles per second, it follows that the light of the sun, 90,000,000 of miles distant, is about eight minutes in reaching us.

THE number of elements or simple substances with which we

are at present acquainted is sixty-two. Of these some 10 or 12 only make up the great bulk of all the objects we see around us.

ONE cubic inch of water, when evaporated, gives about one cubic foot of steam. In other words, water expands, in changing into steam of low temperature, 1,700 times.

ONLY two metals can be welded: Iron and platinum.

To ascertain the specific gravity of a body, find the weight of the body in air and also in water; divide the weight in air by the loss of weight in water, and the result will be the specific gravity.

To give plates of copper a brass color expose the plates, after being sufficiently heated, to the fumes of zinc.

SILICA is the base of the mineral world; carbon the case of the organized.

THE greatest artificial cold, according to Faraday, is 160° Fahrenheit.

A VERY strong glue is formed by throwing a small quantity of powdered chalk into melted common glue.

A GOOD solvent for putty and paint consists of soft-soap mixed with a solution of potash or caustic soda; or pearlash and slackened lime mixed with a sufficient quantity of water to form a paste.

IN a sandy soil the greatest force of a pile-driver will not drive a pile over fifteen feet.

IN order to temper iron sufficiently to cut porphyry, a good plan is to make the iron red-hot, and then plunge it into distilled water from nettles, acanthus, and pilosella, or in the juice pounded out from these plants.

MERCURY freezes at 0.39°.

Iron may be prevented from rusting by first warming it until it is sufficiently hot to burn you, if touched, and then rub it with clean, new white wax. Put it again to the fire until it has soaked in the wax, and afterwards rub well over with a piece of serge.

To make a glue which will resist the action of water, boil one pound of common glue in two quarts of skimmed milk.

An excellent cement for steam-pipes is made of linseed oil varnish ground, with equal weights of white lead, oxide of manganese, and pipe-clay.

In a vacuum water boils at 98° to 100°, according as the vacuum is more or less perfect.

To render paper fire-proof dip it in a solution of alum, and then allow it to dry slowly. To test it, try a slip of the paper in the flame of a candle, and if not sufficiently prepared, dip and dry it a second time.

Paper for copying may be prepared by applying, with a brush, a varnish consisting of Canada balsam and turpentine, equal parts mixed. The paper should be allowed to dry, and then if not transparent enough, the operation should be repeated.

The earth is nearly as heavy as it would be if its mass were entirely composed of metallic antimony or cast-iron.

The temperature of the soil is effected by, first, the exposure of the surface; second, the nature of the soil; third, its permeability by rain, and the presence of underground springs; fourth, the sun's declination; fifth, the elevation above the sea, and, consequently, the heating power of the sun's rays; and, sixth, the amount of cloud and sunshine.—*J. D. Hooker.*

The bark of a dog can be heard at the distance of eighteen hundred yards; the voice of a man at a thousand yards; and the croak of a frog at nine hundred. Taking the difference of size into consideration, the frog has altogether the best lungs.

If Benzole is poured on a piece of ordinary paper immediate transparency is produced to such an extent as to enable one to dispense entirely with tracing paper. On exposure to air, or, better, to a gentle heat, the liquid is entirely dissipated, the paper recovers its opacity, and the original design is found to be quite uninjured.

A ball one foot in diameter just conceals the moon's face when held before it at a distance of one hundred and twenty feet from the eye. Consequently a ball one mile in diameter would do the same thing at a hundred and twenty miles ; a ball of one thousand miles, at one hundred and twenty thousand miles ; and a ball two thousand miles across, at one hundred and twenty times two thousand, or two hundred and forty thousand miles. But this is about the moon's distance, consequently the moon's breadth must be about two thousand miles.

A man weighing one hundred and fifty pounds contains one hundred and eleven pounds of water in his tissues.

The average depth of the Atlantic Ocean is set down at thirteen thousand four hundred feet, and that of the Pacific at eighteen thousand. On the western side of St. Helena soundings have been made, it is said, to the depth of twenty-seven thousand feet — five miles and a quarter — without touching bottom.

A finely polished sheet of iron, three feet long and twelve inches wide, and weighing but three and a half ounces, has been rolled out by a rolling-mill in Pennsylvania. It is thinner than ordinary writing paper.

The Amazon drains an area of two million five hundred thousand square miles ; its mouth is ninety-six miles wide, and it is navigable two thousand two hundred miles from its mouth.

Silver melts at 1873°.

THE sun can never remain in total eclipse at any spot on the terrestial surface for a longer period than three minutes and a quarter of time.

———

THE larger the surface involved the more intense is the feeling of temperature. Water at one hundred and four degrees seems less warm to one finger than water two degrees lower seems to the whole hand.

———

THE first telescope made, — Galileo's feeble instrument, — magnified objects only seven times, and yet with it he discovered the satellites of Jupiter.

———

To clean marble wash the spots with a powder of chalk one part; pumice one part; common soda two parts, mixed; then clean the whole of the stone and wash off with soap and water.

———

A PERMANENT BLACK FOR SILVER.—Having first burnt lead and pulverized it, incorporate it next with sulphur and vinegar to the consistency of a painting color. It is now ready to write with on silver. After applying let it dry, then present it to the fire so as to heat the work a little, and it is finished.

———

THE yearly fall of rain in London is 25 inches; at Vera Cruz, in the Mexican Gulf, rain falls to the amount of 278 inches.

———

A COMPOSITION for bronzing wood, plaster, etc., is made by mixing yellow ochre, Prussian blue, and lamp-black, dissolved in glue-water.

———

PARCHMENT may be closely imitated by dipping unsized paper for about forty seconds in strong sulphuric acid and afterwards in weak ammonia.

———

OIL may be extracted from marble or other stone by making a mixture of one part soap, two parts fuller's earth and one part potash, stirred in with boiling water. Lay it on the spots and allow it to remain a few hours.

THE flash of lightning, by its reflection on the clouds, may be seen from 150 to 200 miles.

TOOTHED wheels; two good rules for making: As the number of teeth in the wheel+2.25 is to the diameter of the wheel, so is the number of teeth in the pinion+1.5 to the diameter of the pinion.

2nd. As the number of teeth in the wheel+2.25 is to the diameter of the wheel, so is (number of teeth in pinion+number of teeth in wheel)÷2, to the distance of their centres.

IN some parts of Egypt it never rains; in Peru, part of Guatemala and part of California but seldom; over the great African Desert, and the desert of Gobi, in Central Asia, and in portions of Arabia and Persia, rain scarce ever falls, while in Guiana, in South America, it rains for a great portion of the year.

PRISMATIC diamond crystals may be formed on windows by applying hot a solution of sulphate of magnesia and a clear solution of gum arabic, well mingled.

MATERIAL OF COMBUSTION.	Lbs. of Water which a ℔. can heat from 0° to 212°.	Lbs. of Boiling Water evaporated by 1 ℔.	Weight of Air at 32° to burn 1 ℔
Ordinary Wood.	26	4.72	4.47
Charcoal	73	13.37	11.46
Turf	30	5.45	4.60
Turf Charcoal	64	11 63	9.86
Coke	65	11.81	11,46
Pit Coal	60	10.90	9 26
Oil. Wax Tallow	78	14.18	15.00
Carburetted Hydrogen	76	13 81	14.58
Alcohol (commercial)	52	9 56	11.60

THE weight of 100 cubic inches of air, in avoirdupois grains (barometer 30, thermometer 62), is 30–49, and the logarithm is 1–48416.

ICE melts at 0.32° Fahrenheit.

A CUBIC inch of water at 62° Fahrenheit, weighs a part of a pound expressed by the decimal 0.036065, and a column of water one foot in height, with a base of one square inch, will, of course, be twelve times this decimal, which is 0.4328. Thus we deduce the following table:

Depth in feet.	Pressure per Square Inch.	Pressure per Square Foot.	Depth in feet.	Pressure per Square Inch.	Pressure per Square Foot.
1	0.4328	62.3232	6	2.5968	373.9392
2	0.8656	124.6464	7	3.0296	436.2624
3	1.2984	186.9696	8	3.4624	498.5856
4	1.7312	249.2928	9	3.8952	560.9088
5	2.1640	311.6160	10	4.2880	623.2320

By the above table the pressure of water on any part of the surface of a receptacle can be ascertained when the depth of such surface is known. If the liquid contained in the vessel be something beside water, it is only necessary to compare the difference in the specific gravity of the two and then calculate as before.

OIL which will not corrode or thicken, and which is much used by watchmakers, is simply prepared by inserting coils of thin sheet lead into a bottle contairing olive oil, and then exposing to the sun and afterwards turning off the clear.

ONE of the best welding mixtures is made as follows: Mix sal ammoniac with ten times its weight of borax. After fusing the mixture pour it on an iron plate. Pulverize when cold, and mix with an equal quantity of quick lime. Apply the powder to the iron when red-hot and the welding will be effected at a lower heat than without the powder.

LEAP YEAR.—Every year the number of which is divisible by 4, without a remainder, is a leap year, except the last year of the century, which is a leap year only when divisible by 400 without a remainder. Thus the year 1900 will not be a leap year.

POWER OF VARIOUS SPECIES OF FUEL.

KINDS OF FUEL.	Effect in ℔s. of water heated 1° by 1 ℔ of fuel.	Effect in ℔s. of water converted into steam of 220°.	Quantity to convert a cubic foot of water into low pressure steam.	Quantity to convert a cubic foot of water into steam, allowing 10 per cent. for loss.
Ordinary Oak	3,600	3.07	20.31	22.6
Oak wood, dry	6,600	5.13	12.2	13.6
Peat, dry and compact	3,250	2.8	22.5	25.0
Coke	9,000	7.7	8.1	9 0
Splint Coal	7,900	6.75	9 25	10.28
Caking Coal	9,800	8.4	7.45	8.22

BRASS ferrules for tool handles, etc., may be perfectly soldered by lapping round the jointing a piece of wire, and then, after wetting, filling the joint with borax and holding in the fire till the brass fuses.

THE worn parts of plated goods can be re-silvered by the following composition: Common salt, 30 gr.; cream of tartar, 3½ gr.; nitrate of silver, 30 gr.; mix; moisten with water and rub over the surface to be plated.

To prevent iron from rusting, rub the surface, after being heated as hot as the hand will bear, with new and clean white wax, and then, after heating again, wipe well with a piece of soft cloth.

To ascertain the length of the day and night at any time of the year, add 12 hours to the time of the sun's setting, and from the sum subtract the time of rising for the *length of the day*. Subtract the time of setting from 12 hours, and to the remainder add the time of rising next morning, for the *length of the night*.

To brass plates of copper, heat them sufficiently and then expose them to the fumes of zinc.

LEAD pencil drawings may be rendered permanent by applying isinglass dissolved in 50 times its weight of water.

TORTOISE-SHELL may be soldered by holding the broken edges in close contact with pinchers heated as hot as they can be without burning the shell. To test this point, try the pinchers on a piece of paper.

To soften ivory, take 15 ounces of water and 3 ounces of spirits of nitre, and let the ivory soak in a closely covered vessel, three or four days after which it will be found to be flexible to the fingers. To harden it again, wrap it in salt for 24 hours.

ABOUT 112 little planets called asteroids have been discovered between the orbits of Mars and Jupiter. Some of them are so small that to call them *worlds* might seem inappropriate. But any body having an *orbit*, though not larger than a cannon ball. is entitled to rank as a world. Many such may exist which will never be discovered because of their littleness. Very little asteroidal bodies are revolving around the earth, as the moon does. The ærolites, or falling stones, are really such asteroids whose orbits cross the earth's orbit, or else wind spirally inward to the earth as a center.

THE following rules will be found useful, as they give the power of water-wheels, with proper allowance for friction and waste of water :

For an undershot : Multiply height of fall by quantity of water flowing per minute, and divide by 5,000. The result will give the number of horse power the effect is equal to.

For an overshot : Multiply the power on an undershot by $2\frac{1}{2}$, and the result will be the number of horse power.

For a breast-wheel : Find the power of an undershot from the top of the fall to where the water enters the bucket : then for an overshot for the rest of the fall. The sum of these two is the power of the breast-wheel.

The quantity of water flowing per minute and the height of the fall are both taken in feet.

Fac Similes of signatures may be neatly and accurately made by sprinkling over the ink before it is dry some finely powdered gum arabic, and then pouring on a fusible alloy in a liquid state. From the plates thus formed copies can be made with printing ink and a copper-plate press.

It is a common error to suppose that many of our greatest inventions and discoveries were made by accident. Many wonderful anecdotes are told in support of this assertion; but a multitude of facts might be adduced to prove that knowledge is more regularly progressive than is commonly imagined. Great discoveries are not made without preparation, and previous knowledge is necessary to turn what are called accidental occurrences to good account.

Hard white metal is made of 1 part tin, 8 parts spelter, and 20 of brass.

Sir Isaac Newton defined mechanics as the geometry of motion.

Distilled water is about 825 times heavier than air.

A machine has been invented with which a writer, using a pen in the usual way can, at the same time, produce a duplicate so small as to be invisible to the naked eye, but so distinct that a microscope will bring out every line and dot. One of the most useful of the applications of this invention will be for preventing forgery, as private marks can be made on notes and securities which will be legible with a microscope, but which an imitator could not see or suspect the presence of.

Gun Cotton is made by immersing cotton-wool in a mixture of sulphuric acid and the strongest nitric acid, or of sulphuric acid and nitrate of potash.

Iceland or calc spar is native carbonate of lime in its primitive crystaline form.

The heat of the human blood is 98°.

PLATINUM, with one exception the heaviest body known, is 21½ times heavier than water.

BATH metal is made of 32 parts brass and 9 parts spelter.

A GOOD boot and shoe polish is made with the whites of two eggs, a table spoonful of alcohol, a lump of sugar, and sufficient finely powered ivory black to thicken.

COMMON bell metal is made of 100 parts of copper and 50 of tin.

A GAS, in expanding to its original volume, after compression, absorbs as much heat as it evolved during the compression.

THE coal product of Great Britain is over 100,000,000 tons. The area of the coal fields lying within the United States has been estimated at 150,000 square miles. The extent of the fields in Ohio alone is not less than 10,000 square miles, or equal to that possessed by Great Britain, and far in excess of that of any other European nation.

THE earth is nearly as heavy as it would be if its mass were entirely composed of metallic antimony or cast-iron.

ROBERT HOOD, an ingenious experimentalist of a long past age, succeeded in making artificial imitations of the circular pits or craters of the moon, by heating strong calcareous solutions until vapor burst out in bubbles through the external surface of the mass.

THE density of the earth's substance has been ascertained by counting the difference in the number of beats a pendulum makes in a given time, according as it is swung on the earth's surface, or down in a deep mine. Mr. Airy, the British astronomer, royal, has made careful experiments of this kind in a coal mine twelve hundred and sixty feet deep. He found that a pendulum which beat seconds at the mouth of the pit made two and a quarter more beats every twenty-four hours when removed to its bottom. The reason is that the pendulum is filled with greater force, and is made to move more rapidly when brought nearer to the earth's center of attraction.

FINE white German silver is made of 1 part iron, 10 parts nickle, and 20 parts copper.

A NEW and effective remedy for burns was accidentally discovered in France in 1869 by a varnisher of metals, who, having got his hand severely burned while at work, and not knowing what to do to deaden the acute pain, thrust his hand into a pot of varnish. The pain ceased as if by enchantment. On the day following he made a further application of the varnish, and in a few days a new skin was formed over the burn, and the hand recovered its wonted flexibility. This simple remedy was afterwards tested by medical men of Paris, and proved to be greatly superior to the usual remedy of nitrate of silver.

TABLE OF THE CORRESPONDING DEGREES ON THE SCALES OF THE
THERMOMETERS OF FAHRENHEIT, REAUMER AND
CENTIGRADE OR CELSIUS.

Fahr.	Reaum.	Cent.	Fahr.	Reaum.	Cent.	Fahr.	Reaum.	Cent.
212	80	100	149	52	65	50	8	10
203	76	95	140	48	60	41	4	5
194	72	90	131	44	55	32	0	0
185	68	85	122	40	50	23	4	5
176	64	80	113	36	45	14	8	10
167	60	75	104	32	40	5	12	15
158	56	70	95	28	35	4	16	20
			86	24	30	13	20	25
			77	20	25	22	24	30
			68	16	20	31	28	35
			59	12	15	40	32	40

THE iron product of the United States is already over 2.000,000 tons, and rapidly increasing.

A CUBIC foot of water weighs 1,000 ounces. To ascertain the weight of a cubic foot of any substance, multiply its specific gravity by 1.000, and the product will be in ounces.

WATER boils at 212°.

THE velocity of a moving body is found by dividing the space passed over by the time consumed in moving over it. Thus, if a body moves 20 miles in 2 hours, its velocity is ascertained by dividing the space 20 by the time 2, which gives a result of 10 miles an hour.

To find the space passed over by a moving body, multiply the velocity by the time.

To find the time occupied by a body in motion, divide the space by the velocity.

SOME years since there was constructed in England a sun-glass, or lens, three feet in diameter, with a focal distance of six feet eight inches. The heat concentrated in the focus of this powerful instrument was sufficient to melt the metals, and even to volatilize them; while flint, quartz, and the most refractory of earthy substances, were easily liquified and caused to boil.

A GOOD varnish for gun-barrels, after browning them, is a dissolved and filtered mixture of 1 oz. shellac, $\frac{1}{4}$ oz. dragon's blood, and 1 qt. rectified spirit.

To remove old paint and putty, lay on with an old brush or rag a solution of caustic soda or potash mixed with soft soap, and allow it to remain a few hours.

BRONZING LIQUID. — Cream of tartar, 3 oz. ; salt, 6 oz. ; sal-ammoniac, 1 oz. Dissolve in a qt. of hot water, then add 2 oz. of nitrate of copper dissolved in $\frac{1}{2}$ pt. of water.

MODELING WAX is made of white wax with sufficient lard added to make it work easy. To prevent adhering in use the tools and board should be moistened with water.

A LIQUID AMALGAM for silvering globes, etc., is made of 1 oz. pure lead, 1 oz. grain tin, which being melted in a clean ladle, add immediately 1 oz. of bismuth. After skimming off the dross remove the ladle from the fire, and before the metal sets add 10 oz. of quicksilver. Mix well, and avoid the fumes.

Drying by Vacuum.—As the juices contained in all vegetable substances are retained in their particular vessels by the pressure of the atmosphere, by removing that pressure the fluids can no longer be retained by the veins and recepticles. This principle has been effectively applied, among other uses, to the drying of lumber. An air-tight reservoir, provided with one or more air pumps, is filled with the green lumber. The air pump is then set in motion, and heat at the same time applied to the exterior of the reservoir. As the pressure of the atmosphere is removed the vegetable juices begin to exude from the pores of the wood, and being vaporized by the heat, are carried off with the air through the pump. Wood and other substances are dried very rapidly by this process, and it is deserving of more attention than has hitherto been given to it.

For cleaning varnish use weak ley of potash or soda mixed with a little powdered chalk.

MECHANICAL POWERS.

The lever may be considered the first, and, in fact, the foundation, of all mechanical powers, as the effects produced by all of them are ultimately resolvable into that of the lever.

They are divided into three classes.

First. Those in which the fulcrum is situated between the weight and the power.

Second. Those in which the weight is situated between the fulcrum and the power.

Third. Those in which the power is situated between the weight and the fulcrum.

The bent lever is of the first class, and has no particular advantage except that of form, which is given to it for convenience in use.

To preserve an equilibrium between the power and the weight, they must be to each other inversely as their distances from the fulcrum.

The following are the rules for calculating the effect of the several levers:

Class 1, or when the fulcrum is between the power and the weight.

Rule: Divide the weight to be raised by the power to be applied; the quotient will give the difference of leverage necessary to support the weight in equilibrio. Hence, a small addition, either of leverage or weight, will cause the power to preponderate.

Example: Five men, who exert a combined force of 15 cwt., desire to raise a weight of six tons. Required the proportionate length of lever.

$$6 \text{ T.} = 120 \text{ cwt.} ; \text{ and } \frac{120}{15} = 8$$

In this example, the proportionate lengths of the lever to maintain the weight in equilibrio are as 8 to 1.

Class 2. When the fulcrum is at one end of the lever, and the power at the other, with the weight between.

Rule: As the distance between the power and the fulcrum is to the distance between the weight and fulcrum, so is the effect to the power.

Example: It is necessary to raise 240 ℔s. when the weight is placed 8 feet from the power and 2 feet from the fulcrum.

$$6:2::240:80.$$

THE PULLEY.—Pulleys are known as fixed and moveable.

There is merely a change in the direction, and no economy of power in the use of a fixed pulley. A saving equal to one-half the power is effected by the employment of the moveable pulley, since it changes its position with that of the weight. When the weight is equal to the product of the power, and twice the number of moveable pulleys, an equilibrium is attained between the weight and the power.

To find the power necessary to raise a given weight through the medium of a system of moveable pulleys.

Rule: Divide the weight to be raised by twice the number of pulleys in the lower block; the quotient will give the power necessary to raise the weight.

Example: Required the power to raise 800 ℔s. — the lower block containing four pulleys.

$$\frac{800}{4 \times 2} = 100 \text{ ℔s.}$$

WHEEL AND AXLE.—This device may be looked upon as simply a revolving lever. An equilibrium between the power acting on the circumference of the wheel, and the weight or resistance acting on the circumference of the axle, is attained when the power is to the weight as the radius of the axle is to that of the wheel.

To ascertain the effective gain the following rule may be employed:

Rule: As the radius of the wheel is to the radius of the axle, so is the effect to the power.

Example: There is exerted on the periphery of a wheel whose radius is eight feet a weight of sixty pounds. Required the weight raised at the extremity of a cord wound round the axle, the radius being twelve inches.

$$\frac{60 \text{ ℔s.} \times 8 \text{ feet}}{12 \text{ inches.}} = 480$$

INCLINED PLANE.—The inclined plane becomes a mechanical power in consequence of its supporting a part of the weight and of course leaving only a part to be supported by the power. The direction of the force is changed from the perpendicular to one more or less horizontal, and the weight moves upward on it in a diagonal between them. When the power is to the weight as the perpendicular height of the inclined plane is to its hypothenuse, equilibrium is attained; but as the height is to the base when in a direct parallel to the base.

Rule: As the length of the plane is to its height so is the weight to the power.

Example: Required the power necessary to raise 900 ℔s. up an inclined plane 6 feet long and 2 feet high.

As 6:2::900:300 ℔s. the power.

WEDGE.—The wedge is simply a combination of two inclined planes united together along their bases, and made to act upon two weights or resistances at once, or on a fulcrum and a weight between which it is caused to move by power applied to its base, generally, in practice, by the impulse of successve blows.

Equilibrium is attained when the power is to the resistance as the base of the wedge is to its length, or to the length of its side, accordingly as the resistance acts perpendicularly to the central line of lenghth, or to that of the side.

Case 1. When a wedge is employed to force two bodies from one another, in a direction parallel to the base of the wedge.

Rule: As the length of the wedge is to half its back or head, so is the resistance to the power.

Example: Required the power necessary to separate two substances with a resistance of 150 ℔s., — the breadth of the back or head of the wedge employed being 4 inches, and the length of either of its inclined sides 12 inches.

As 12:2::150:25 ℔s. the power.

Case 2. When only one of the bodies is moveable.

Rule: As the length of the wedge is to the breadth of the back or head, so is the resistance to the power.

Example: The breadth of the back or head being 5 inches, and the length of either of the inclined sides 15 inches, required, the power necessary to separate two substances with a resistance of 225.

As 15:5::250:75 ℔s. the power.

SCREW.—The screw is simply a modification of the inclined plane, and may be considered as a simple cylinder, having wound, spirally, around its periphery an inclined plane in the form of a triangle. The circumference of the cylinder forms the base of the triangle; its hypothenuse forms the spiral cord or inclined plane; and the distance between two consecutive cords or threads, the height of the incline.

Rule: To the square of the circumference of the screw add the square of the distance between two threads, and extract the square root of the sum; this will give the length of the inclined plane. Its height is the distance between two consecutive cords or threads.

Where a lever or wrench is employed to turn the screw the power of the screw is as the circle described by the handle of the wrench or lever, to the internal or distance between the spirals.

Case 1. When the weight to be raised is known, to find the power.

Rule: Multiply the weight by the distance between two threads of the screw, and divide the product by the circumference of the circle described by the lever. The quotient is the power.

Example: Required the power to be applied to the end of a lever 5 feet long, to raise a weight of 8 tons.

$$\frac{16,000 \times 1\frac{1}{2}}{60 \text{ inches} \times 2 \times 3, 1416} = 63,664.$$

MECHANICAL POWERS.—All mechanical powers or machines are simply combinations of one or more of the six elementary mechanical powers, viz.: the lever, the pulley, the wheel and axle, the inclined plane, the wedge, and the screw; and the mechanical effects of the whole are ultimately resolvable into that of the lever.

Although there is a gain of power in the use of the mechanical powers it is only at the sacrifice of a corresponding loss of time.

A great resistance may be overcome slowly, or a great weight may be sustained, by the application of a small force; or, a small weight or resistance may have a great velocity imparted to it by the employment of a great force or power.

POPULATION OF CITIES.

CITIES.	1870.	1860.	1850.
New York, N. Y.	926,341	813.669	515,547
Philadelphia, Penn.	674,022	565,529	408.762
Brooklyn, N. Y.	396,800	266.661	96,837
St. Louis, Mo	312,963	212,418	77,860
Chicago, Ill	298,983	109,260	29 963
Baltimore, Md.	267,354	212.418	169,054
Boston, Mass.	250,526	177.841	136,881
Cincinnati, Ohio	216,239	161,044	115,436
New Orleans, La	191,322	168.675	116,375
San Francisco, Cal	149,482	56,802	34,870
Buffalo, N. Y.	117,715	81,129	42,261
Washington, D. C.	109,204	61,122	40,001
Newark, N. J.	105,078	71,941	38,894
Louisville, Ky.	100,754	68.033	43.194
Cleveland, Ohio.	92,846	43,417	17,034
Pittsburgh, Pa.	86,235	49,217	46,601
Jersey City, N. J.	81,744	29,226	6,856
Detroit, Mich.	79,580	45.619	21.019
Milwaukee, Wis.	71.499	45.246	20.061
Albany, N. Y.	69,422	62,367	50,763
Providence, R. I.	68,906	50.666	41,513
Rochester, N. Y.	62,385	48.204	36,403
Alleghany City. Pa.	53.181	28,702	....
Richmond, Va.	51.038	37.910	27 570
New Haven, Conn	50,840	39.267	20.345
Charleston, S. C.	48,956	40,522	42,985
Troy, N. Y.	46.471	39,235	28,785
Syracuse, N. Y.	43 051	28.119	22.271
Worcester, Mass.	41.105	24.960	17.049
Lowell, Mass.	40,928	36.827	33 383
Memphis, Tenn.	40,226	22,623	8.839
Cambridge, Mass.	39.634	26,060	15 215
Hartford, Conn.	37,180	29,152	13.555
Indianapolis, Ind.	36.565	18,611	8,034
Scranton, Pa.	35,093	9,223	....
Reading, Pa.	33 932	23,162	....
Columbus, Ohio	33.745	18,554	17,882
Paterson, N. J.	33.582	19,586	11 334
Kansas City, Mo.	32.260	....	....
Dayton, Ohio.	32,579	20 081	10.977
Mobile, Ala	32,084	29.258	20,515
Portland, Me.	31.414	26,341	20.815
Wilmington, Del.	30,841	21,258	13 979
Lawrence, Mass	28,921	17 639	8.282
Toledo, Ohio.	28 546	13.768	3 829
Charlestown, Mass.	28,323	25.065	17.216
Lynn, Mass.	28,233	19.083	14.257
Fall River. Mass.	26 780	14,026	11,524
Springfield, Mass.	26,703	15,199	11,766
Nashville, Tenn	25.872	16 988	10,165
Peoria, Ill.	25 787	14,045	....
Covington, Ky	24,505	....	....
Salem, Mass.	24,117	22,252	20,264
Quincy, Mass.	24,053	....	....

POPULATION OF CITIES—*Continued.*

CITIES.	1870.	1860.	4850.
Manchester, N. H..	23,536	20,107	....
Harrisburg, Pa..	23,109	13,405	7,834
Trenton, N. J..	22,874	....	....
Evansville, Ind..	22,830	11,484	3,235
New Bedford, Mass..	21,820	22,300	16,443
Oswego, N. Y..	20,910	16,816	12,205
Elizabeth, N. J..	20,838	....	....
Lancaster, Pa..	20,233	17,603	12,369
Savannah, Ga..	20,233	....	....
Camden, N. J..	20,045	....	....
Davenport, Iowa..	20,042	11,267	....
St. Paul, Minn..	20,081	10,401	....
Wheeling, Va..	19 282	....	....
Norfolk, Va..	19,256	14,620	14,326
Taunton, Mass..	18,629	15,376	....
Chelsea, Mass..	18,547	13,395	....
Dubuque, Iowa..	18 404	....	....
Leavenworth, Kan..	17,849	7,429	....
Fort Wayne, Ind..	17,718	....	....
Springfield, Ill..	17,365	....	....
Atlanta, Ga..	16 988	....	....
Norwich, Conn..	16,653	14,048	....
Sacramento, Cal..	16,484	13,785	....
Omaha, Neb..	16,083	....	....
Gloucester, Mass..	15 389	10,904	....
New Brunswick, N. J..	15,059	....	....
New Albany, Ind..	14,973	12,647	9,895
Galveston, Texas..	13,818	7,307	....
Newburyport, Mass..	13,595	13,401	9,572
Alexandria, Va..	13,570	....	....
Wilmington, N. C..	13,446	....	....
Newport, R. I..	12,521	10,508	....
Little Rock, Ark..	12,380	....	....
Concord, N. H..	12,241	....	....
Des Moines, Iowa..	12 035	....	....
Waterbury, Conn..	10,826	10,004	....
Nashua, N. H..	10,543	10,065	....
Raleigh. N. C..	10,149	....	....
New London, Conn..	9 576	10 115	....
Portland, Oreg..	8,293	2,874	....
Virginia City, Nev..	7,008	....	....
Topeka, Kan..	5,790	....	....

POPULATION OF THE STATES OF THE UNION.

STATES. (37)	AREA. Square Miles.	Total Population 1870.	Increase from 1860 to 1870.	Increase Per Cent
Alabama	50,722	1,002,000	37,799	3.92
Arkansas	52,198	473 174	37,724	8 66
California	188,981	549,808	169 814	44.69
Connecticut	4,750	537,417	77,270	16.79
Delaware	2,120	125,015	12,799	11.41
Florida	59 248	189.995	49,571	35,30
Georgia	58,000	1,174,832	117.550	11.12
Illinois	55,410	2,529.410	817,459	47.75
Indiana	33,809	1,655 675	305,247	22.60
Iowa	55,045	1,181 359	506,446	75.04
Kansas	81,318	379,497	272 291	253 99
Kentucky	37,680	1,320,407	164.723	14.99
Louisiana	41,346	734 420	26,418	3.73
Maine	35,000	628,719	440	.07
Maryland	11,124	790.095	103.046	15 00
Massachusetts	7,800	1,457,351	226,285	18.38
Michigan	56,451	1,184,653	435 540	58.14
Minnesota	83,531	424,543	252,520	146.79
Mississippi	47.156	842 056	50,751	6.41
Missouri	65 350	1,691,693	509.681	3.03
Nebraska	75.995	116,888	116,888	305.28
Nevada	81,539	42 456	42,456	519 16
New Hampshire	9,280	317,710	8,633	†2.56
New Jersey	8,320	903.044	231.009	34.37
New York	47,000	4,370,846	850,111	12.63
North Carolina	50,704	1,016,954	24,332	2.45
Ohio	39.964	2,652,302	312.791	13.33
Oregon	95,274	90 878	38.413	73.64
Pennsylvania	46 000	3,511,543	605 328	20 83
Rhode Island	1 306	217.356	42,736	24.46
South Carolina	34,000	705.789	2,081	.30
Tennessee	45,600	1,225 937	116,136	10.46
Texas	274,356	800.000	195,785	32.46
Vermont	10,212	330,582	15,484	4.91
Virginia	38,352	1,211.442	56,218	3.52
West Virginia	23,000	441.094		
Wisconsin	53,924	1,055,501	279,621	36.04

FOREIGN PATENTS.

AMERICANS who have inventions of importance and value for which they anticipate taking out patents in this and foreign countries, cannot be too cautious of the manner in which they proceed, and we would advise all such to procure experience and advice before allowing themselves to take the first step toward applying even for a patent in the United States, otherwise, owing to the peculiarity of foreign patent laws, they may unwittingly defeat the very object they have in view. In some foreign counties, the *first importer* can procure a patent, and there is therefore, no absolute safety for the actual inventor except by making his application in certain countries abroad, prior to the issue of his U. S. patent.

That Europe offers many rich and important fields to the inventor no one will deny. England we need hardly refer to, for despite the peculiarity of her patent laws, she has long offered to inventors a field second in importance, if not equal to the United States. The forms and technicalities of the British law and practice, necessitate unusual caution and experience in applying for patents to avoid defeat, and inventors (especially foreign) have long hoped for a chance more liberal. France, despite the trouble through which she has recently passed, still offers a field hardly less advantageous than that of England. Belgium, though one of the least, is one of the most important of European States to command the attention of patentees and manufacturers. But we wish particularly to call American inventors and patentees' attention to the splendid and rapidly improving fields of Austria, Prussia and Russia. The introduction of new inventions into these countries has been too long neglected by foreign patentees and owners of valuable patents. Russia especially, absolutely requires agricultural and other labor-saving machinery to develop her vast natural resources. Austria readily adopts useful inventions, and Prussia is undoubtedly the most advanced and enterprising state of Continental Europe.

Though we would not advise inventors to secure patents in the colonies indiscriminately, as a large majority of inventions of unquestionable value in larger lands, are not adapted to the conditions and wants of the colonists, still there are many that could not fail to amply reward the patentee for the expense of introduction.

The risk involved in applying for foreign patents, is, in most instances comparatively small, as examinations as to novelty are seldom made, and the invention is looked upon with favor.

Although we are thoroughly conversant with the numerous difficulties and technicalities to be overcome in prosecuting applications for patents abroad, we feel no hesitation in offering our services in obtaining British, Continental and Colonial patents; and being acquainted with all the modern improvements of importance, and having an extensive practical experience in all the principal classes of machinery of modern times, and in developing inventions to be applied to the useful arts, we are enabled to advise as to the novelty of inventions, and clearly distinguish the old from the new portions of an improvement, which is of great importance in the preparation of specifications for patents.

The following information respecting foreign patents, is from necessity short, but we shall be pleased to furnish further information to inventors and others interested, without charge, and we would say that being connected with reliable and competent agencies in London, Paris, Berlin, St. Petersburg, Vienna and the other important Continental capitals, we continually and promptly receive notice of any changes in the laws or practice of the different countries.

ENGLAND.

ENGLAND is undoubtedly the first and most important country of Europe in which to secure the monopoly of an invention. Patents are granted for the " United Kingdom of Great Britain " for the term of fourteen years, for any *"new manufacture,"* or *" improvement therein,* that is *"new in the realm."* The true inventor, or *first importer* of an invention (citizen or foreign), may obtain such patent. This latter provision renders it particularly important that American inventors should apply for English patents if possible *before,* or as soon after their United States patent is issued as practicable, if they would insure the invention against piracy. Married women, minors, executors or administrators, native or foreign, enjoy like privileges. Great care and vigilance is necessary in the preparation and prosecution of applications, owing to the technicalities of the official practice, and neglect for a single day, or a single error is sufficient to defeat success. Inventors cannot, therefore, be too careful in the selection of experienced and reliable attorneys. Space forbids our entering into details, as the subject would occupy a volume itself, but we shall be pleased to answer all inquiries without charge. Being immediately connected with one of the oldest and most reliable agencies in London, we are able to give each case entrusted to our hands, constant and personal attention before the office,—a necessity almost imperative to success, but possessed by few agents in this country.

FRANCE.

THE law requiring the patentee to be the inventor or author, is now liberally interpreted, and it is held that the proprietor or foreign patentee, though not the actual inventor, may obtain a valid patent, and a firm or company secure a patent in its corporation or firm name. Patents are granted for fifteen years, unless the invention has been previously patented abroad, in which case it would bear date and expire with the previously

obtained patent. Two years time is allowed in which to introduce and work the patent, and in order to keep the patent
alive, small annuities are required. French patents cover the
colonies.

All changes in foreign patent laws are promptly reported to
us, so that we can at all times furnish correspondents with the
latest information.

BELGIUM.

THIS country, though small in comparison to her neighbors,
is by no means the least of importance to patentees. She
presents a rich field for the working and disposing of useful
inventions, on account of the advanced state of the useful arts
to which her thick population have attained. Patents are granted
for twenty years. Besides the first expense of obtaining a
patent, a yearly government tax has to be paid to keep it in
force. It is also necessary that the patent should be worked
within a year of its introduction in any other country ; and if
the invention has been patented elsewhere, it expired with
such foreign patent.

These requirements render it necessary that inventors should
exercise caution in introducing their inventions abroad, if they
wish to protect themselves in Belgium, and we would advise
them to consult competent attorneys before making their invention too public.

Applications for patents for Belgium have to be made in the
French language, with great care, and we offer patentees not
only our services and superior facilities at a reasonable expense,
but we will readily give them advice on the subject gratis.

AUSTRIA.

ANOTHER fine field for the inventor and patentee is Austria.
Patents may be obtained by the actual inventor or his assignee,
or authorized agent—native or foreign,—for any ''discovery,
invention, or improvement,'' not known or used within the

empire, for a term of fifteen years or less. Preparations of food, medicines and beverages, or inventions contrary to sanitary laws or public regulations, are not patentable. The existence of a foreign patent would not prevent a patent issuing upon the same invention, provided it has not been worked within the empire, but such patent would be limited in duration to that of the patent abroad. A patent for less than fifteen years may be prolonged, but an extension beyond fifteen years can only be obtained by a special decree of the Emperor. It is necessary to work and introduce the invention within one year of the date of issue. Minor details and advice free by applying to us.

HUNGARY.

A SEPARATE patent is now granted the Kingdom of Hungary, and additional drawings and papers are required, but the cost is included in the patent for Austria, and all the regulations are the same.

RUSSIA.

PATENTS in this country are growing of more value and importance every year. In 1869, the government imposed a duty upon foreign machinery which had previously been admitted to the country free, excepting only agricultural machines and those for working fibrous materials. As a consequence of the enforcement of this duty, combined with rapid increase of railroads throughout the country, home-manufactures are rapidly increasing both in demand and production, and inventions are rapidly disposed of. This is especially the case with labor-saving inventions, and those relating to railroads, agriculture, and the manufacture of iron and steel.

Although the first cost of a patent in Russia is more than in any other European state, we believe for many inventions, the right in this country is proportionately the most valuable.

Patents are allowed for three, five, and ten years, at the choice of the petitioner, who may be either the inventor or his assignee,—a native or foreigner. Further particulars, costs, etc., furnished by us free of charge.

SPAIN.

THE law denying an inventor a patent on firearms or other munitions of war has been modified, and inventors can now have full protection in this country in all branches of industry. When the invention is an importation, it is of five years duration, with the privilege of prolongation to ten years. The invention must be invariably worked within a year from the date of the patent, and must not remain unworked for the same length of time, to be kept in force.

CUBA.

THE Governor-General is allowed to grant patents under the same regulations as those of Spain; but applications are attended with much trouble, and great care must be observed in the preparation.

TURKEY.

"FIRMANS" or special privileges are granted by the Turkish Government for a term varying from three to twenty years, for new inventions. The application must be made in the name of the inventor, or the foreign patentee, if any; otherwise the importer who has the written consent of the inventor may obtain the concession. Military weapons are not patentable. The cost of obtaining the "firman" cannot be fixed, as the expenses depend upon the demands of the officials, no fee being named by the government. We will give further information to correspondents in respect to the probable cost of this patent in any particular case, if they will inform us of its nature and details.

PRUSSIA.

CONTRARY to the general rule in foreign countries, Prussia subjects all applications to an examination as to novelty, which is made by the Royal Polytechnic Commission. If allowed (which is generally the case), the invention has to be introduced and worked within six months; but, if good and sufficient reasons can be shown, an additional six months is allowed. Patents run for from five to sixteen years. Prolongations are not allowed. Americans cannot obtain these patents in their own names, though some other foreigners may under treaty stipulation. For this reason, patents are usually obtained through a resident, and are good and valid, if the matter is attended to with proper care. Our connections with agents in Europe, is such that we have never failed in securing an application that has been entrusted to our hands, and the patents in every instance have been valid and sufficient. Where the invention is upon a sewing-machine, a model is required, but in no other case.

Patents in Prussia, as in most foreign countries, should be applied for before they issue in the United States.

SAXONY.

PATENTS are granted for from five to ten years, and must be worked within one year from the date of the grant. The patent must be secured in the name of a native, or person naturalized in Saxony in trust for the inventor, and can afterwards only be held by a citizen of the German Confederation.

HOLLAND.

PATENTS may be secured in the name of the person possessing the invention, whether the inventor or not, or in the name of a resident; but in case the patentee afterwards secures a patent for the same invention elsewhere, the patent in Holland becomes void. Unless, therefore, the inventor secures this patent last, he should secure it in the name of a resident agent. Patents may be obtained for five, ten or fifteen years, the government tax increasing with the length of the patent. The invention should be worked within two years.

BAVARIA.

IMPORTED patents expire with those previously obtained abroad; otherwise the duration of the patent is fifteen years. A lesser term may be applied for and prolonged from year to year till the fifteen years are attained. Patents ought to be worked within three years, if not imported; in which latter case the the invention must be introduced within one year.

WURTEMBURG.

THE duration of imported patents is the same as those previously obtained abroad; otherwise patents are granted for from one to ten years. The invention is required to be introduced and worked within the first two years.

SWEDEN.

PATENTS for inventions that have been patented abroad expire with the foreign patent. Patents are issued for terms varying between from three to fifteen years. In this country two years are allowed for the introduction and working of a patented invention.

NORWAY.

PATENTS are granted for from five to ten years, the government reserving the right to fix the duration beyond five years. As in the case of Sweden, the invention must be worked within two years from the date of the patent.

PORTUGAL.

IMPORTED patents expire with the previously obtained foreign patent for the same invention, if any; otherwise patents are granted for fifteen years. The law requires the invention to be introduced before the expiration of half the time for which the patent is granted.

DENMARK.

No special legislation exists in the country relative to patents, but special privileges are frequently granted by the crown on the recommendation of the Board of Customs, according to certain established rules. The duration of such patent is fixed by the government in each case, varying from between three to twenty years. The patent should be worked within the first year ot its existence, and is only kept in force by continuous working.

GERMAN PRINCIPALITIES.

In the principalities, no special legislation exists relating to patents, but they are granted by the government of each state on the favorable report of a scientific commission, for periods varying from one to fifteen years.

ITALY.

Patents are allowed for fifteen years, but expire with any previously obtained foreign patent. Additional improvements are allowed, and have the force of the original patent.

GREECE.

We are not aware of any special patent laws existing in Greece, but the Government is empowered to grant Letters Patent for inventions, subject to the approval of the Senate. The practice is uncertain, and the cost variable.

COLONIAL PATENTS.

EAST INDIA.

In this colony the patent laws are excellent, and the exclusive right is obtainable for fourteen years, at a comparatively small cost. Patents issued to the actual inventor or his assignee, and to a foreign inventor or his authorized agent. Patents obtained in England, and afterwards secured in India, will expire with the English patent.

NEW SOUTH WALES, QUEENSLAND AND SOUTH AUSTRALIA.

THESE three colonies are almost identical in their patent laws and practice. Patents issued for from seven to fourteen years to the inventor or his authorized agent. Unlike most foreign countries, these colonies subject the invention to an examination, as to novelty and utility.

NEW ZEALAND.

PROVIDED a patent has not been obtained in the colony or any other country for the same invention, a patent can be obtained for the term of fourteen years, by colonist or alien. The law allows additional improvements and reissues, but no provision is made for extensions, nor is it required to introduce the invention within a certain period.

CEYLON.

THE actual inventor or first importer (native or foreign), of an invention can procure a valid patent for fourteen years, with the privilege of extension for fourteen years additional. Where the invention has been secured in England, it may be protected by filing the specification and certified copy of the English patent, but the protection will expire with such patent.

VICTORIA.

THE law in this colony provides for the patenting of "any new and useful manufacture" by the "true and first inventor," either colonist or non-resident, a mere importer being excluded. The application is subjected to a scientific examination, to determine its novelty or utility, and the rule is enforced with rigor in the case of the application of a foreign inventor, which renders it necessary to secure experienced assistance in the preparation of the application. The patent is good for fourteen years, and the working of the invention is not required within a certain period.

TASMANIA.

EXCEPTING the subjecting of applications to a scientific examination, the laws and practice of this colony are essentially the same as those of Victoria.

THE
PATENT LAWS

OF THE

UNITED STATES OF AMERICA

PASSED JULY 8, 1870.

AN ACT to revise, consolidate, and amend the statutes relating to patents and copyrights.

Be it enacted by the Senate and House of Representatives of the United States of America in Congress assembled, That there shall be attached to the department of the interior, the office heretofore established, known as the Patent Office, wherein all records, books, models, drawings, specifications, and other papers and things pertaining to patents shall be safely kept and preserved.

SECTION 2. *And be it further enacted,* That the officers and employees of said office shall continue to be one commissioner of patents, one assistant commissioner, and three examiners-in-chief, to be appointed by the president, and by and with the advice and consent of the senate; one chief clerk, one examiner in charge of interferences, twenty-two principal examiners, twenty-two first assistant examiners, twenty-two second assistant examiners, one librarian, one machinist, five clerks of class four, six clerks of class three, fifty clerks of class two, forty-five clerks of class one, and one messenger and purchasing clerk, all of whom shall be appointed by the secretary of the interior, upon nomination of the commissioner of patents.

SEC. 3. *And be it further enacted,* That the secretary of the interior may also appoint, upon like nomination, such additional clerks of classes two and one, and of lower grades, copyists of drawings, female copyists, skilled laborers, and watchmen, as may be from time to time appropriated for by congress.

Sec. 4. *And be it further enacted,* That the annual salaries of the officers and employees of the Patent Office shall be as follows :

Of the commissioner of patents, four thousand five hundred dollars.

Of the assistant commissioner, three thousand dollars.

Of the examiners-in-chief, three thousand dollars each.

Of the chief clerk, two thousand five hundred dollars.

Of the examiner in charge of interferences, two thousand five hundred dollars.

Of the principal examiners, two thousand five hundred dollars each.

Of the first assistant examiners, one thousand eight hundred dollars each.

Of the second assistant examiners, one thousand six hundred dollars each.

Of the librarian, one thousand eight hundred dollars.

Of the machinist, one thousand six hundred dollars.

Of the clerks of class four, one thousand eight hundred dollars each.

Of the clerks of class three, one thousand six hundred dollars each.

Of the clerks of class two, one thousand four hundred dollars each.

Of the clerks of class one, one thousand two hundred dollars each.

Of the messenger and purchasing clerk, one thousand dollars.

Of laborers and watchmen, seven hundred and twenty dollars each.

Of the additional clerks, copyists of drawings, female copyists, and skilled laborers, such rates as may be fixed by the acts making appropriations for them.

Sec. 5. *And be it further enacted,* That all officers and employees of the Patent Office shall, before entering upon their duties, make oath or affirmation truly and faithfully to execute the trusts committed to them.

SEC. 6. *And be it further enacted*, That the commissioner and chief clerk, before entering upon ther duties, shall severally give bond, with sureties, to the treasurer of the United States, the former in the sum of ten thousand dollars, and the latter in the sum of five thousand dollars, conditioned for the faithful discharge of their duties, and that they will render to the proper officers of the treasury a true account of all money received by virtue of their office.

SEC. 7. *And be it further enacted*, That it shall be the duty of the commissioner, under the direction of the secretary of the interior, to superintend or perform all the duties respecting the granting and issuing of patents which herein are, or may hereafter be, by law directed to be done; and he shall have charge of all books, records, papers, models, machines, and other things belonging to said office.

SEC. 8. *And be it further enacted*, That the commissioner may send and receive by mail, free of postage, letters, printed matter, and packages relating to the business of his office, including Patent Office reports.

SEC. 9. *And be it further enacted*, That the commissioner shall lay before congress, in the month of January, annually, a report giving a detailed statement of all moneys received for patents, for copies of records or drawings, or from any other source whatever; a detailed statement of all expenditures for contingent and miscellaneous expenses; a list of all patents which were granted during the preceding year, designating under proper heads the subjects of such patents; an alphabetical list of the patentees, with their places of residence; a list of all patents which have been extended during the year; and such other information of the condition of the Patent Office as may be useful to congress or the public.

SEC. 10. *And be it further enacted*, That the examiners-in-chief shall be persons of competent legal knowledge and scientific ability, whose duty it shall be, on the written petition of the appellant, to revise and determine upon the validity of the adverse decisions of examiners upon applications for patents, and for re-issues of patents, and in interference cases ; and when required by the commissioner, they shall hear and report upon claims for extensions, and perform such other like duties as he may assign them.

SEC. 11. *And be it further enacted,* That in case of the death, resignation, absence, or sickness of the commissioner, his duties shall devolve upon the assistant commissioner until a successor shall be appointed, or such absence or sickness shall cease.

SEC. 12. *And be it further enacted,* That the commissioner shall cause a seal to be provided for said office, with such device as the President may approve, with which all records or papers issued from said office, to be used in evidence, shall be authenticated.

SEC. 13. *And be it further enacted,* That the commissioner shall cause to be classified and arranged in suitable cases, in the rooms and galleries provided for that purpose, the models, specimens of composition, fabrics, manufactures, works of art, and designs, which have been or shall be deposited in said office; and said rooms and galleries shall be kept open during suitable hours for public inspection.

SEC. 14. *And be it further enacted,* That the commissioner may restore to the respective applicants such of the models belonging to rejected applications as he shall not think necessary to be preserved, or he may sell or otherwise dispose of them after the application has been finally rejected for one year, paying the proceeds into the treasury, as other patent moneys are directed to be paid.

SEC. 15. *And be it further enacted,* That there shall be purchased for the use of said office, a library of such scientific works and periodicals, both foreign and American, as may aid the officers in the discharge of their duties, not exceeding the amount annually appropriated by congress for that purpose.

SEC. 16. *And be it further enacted,* That all officers and employees of the Patent Office shall be incapable, during the period for which they shall hold their appointments, to acquire or take, directly or indirectly, except by inheritance or bequest, any right or interest in any patent issued by said office.

SEC. 17. *And be it further enacted*, That for gross misconduct the commissioner may refuse to recognize any person as a patent agent, either generally or in any particular case ; but the reasons for such refusal shall be duly recorded, and be subject to the approval of the secretary of the interior.

SEC. 18. *And be it further enacted,* That the commissioner may require all papers filed in the Patent Office, if not correctly, legibly, and clearly written, to be printed at the cost of the party filing them.

SEC. 19. *And be it further enacted*, That the commissioner, subject to the approval of the secretary of the interior, may from time to time establish rules and regulations, not inconsistent with law, for the conduct of proceedings in the Patent Office.

SEC. 20. *And be it further enacted*, That the commissioner may print, or cause to be printed, copies of the specifications of all letters-patent, and of the drawings of the same, and copies of the claims of current issues, and copies of such laws, decisions, rules, regulations, and circulars as may be necessary for the information of the public.

SEC. 21. *And be it further enacted*, That all patents shall be issued in the name of the United States of America, under the seal of the Patent Office, and shall be signed by the secretary of the interior, and countersigned by the commissioner, and they shall be recorded, together with the specification, in said office, in books to be kept for that purpose.

SEC. 22. *And be it further enacted*, That every patent shall contain a short title or description of the invention or discovery, correctly indicating its nature and design, and grant to the patentee, his heirs or assigns, for the term of seventeen years, the exclusive right to make, use and vend the said invention or discovery throughout the United States and the territories thereof, referring to the specification for the particulars thereof ; and a copy of said specifications and of the drawings shall be annexed to the patent and be a part thereof.

SEC. 23. *And be it further enacted*, That every patent shall date as of a day not later than six months from the time at which it was passed and allowed, and notice thereof was sent to the applicant or his agent ; and if the final fee shall not be paid within that period, the patent shall be withheld.

SEC. 24. *And be it further enacted*, That any person who has invented or discovered any new and useful art, machine, manufacture, or composition of matter, or any new and useful improvement thereof, not known or used by others in this country, and not patented or described in any printed publication in this or in any foreign country, before his invention or discovery thereof, and not in public use or on sale for more than two years prior to his application, unless the same is proved to have been abandoned, may. upon payment of the duty required by law, and other due proceedings had, obtain a patent therefor.

SEC. 25. *And be it further enacted*, That no person shall be debarred from receiving a patent for his invention or discovery, nor shall any patent be declared invalid, by reason of its having been first patented or caused to be patented in a foreign country; provided the same shall not have been introduced into public use in the United States for more than two years prior to the application, and that the patent shall expire at the same time with the foreign patent, or, if there be more than one, at the same time with the one having the shortest term; but in no case shall it be in force more than seventeen years.

SEC. 26. *And be it further enacted*, That before any inventor or discoverer shall receive a patent for his invention or discovery, he shall make application therefor, in writing, to the commissioner, and shall file in the Patent Office a written description of the same, and of the manner and process of making, constructing, compounding, and using it, in such full, clear, concise, and exact terms as to enable any person skilled in the art or science to which it appertains, or with which it is most nearly connected, to make, construct, compound, and use the same: and in case of a machine, he shall explain the principle thereof, and the best mode in which he has contemplated applying that principle, so as to distinguish it from other inventions; and he shall particularly point out and distinctly claim the part, improvement, or combination which he claims as his invention or discovery; and said specification and claim shall be signed by the inventor and attested by two witnesses.

SEC. 27. *And be it further enacted*, That when the nature of the case admits of drawings, the applicant shall furnish one copy signed by the inventor or his attorney in fact, and attested by two witnesses, which shall be filed in the Patent Office; and a copy of said drawings, to be furnished by the Patent Office, shall be attached to the patent as a part of the specification.

SEC. 28. *And be it further enacted*, That when the invention or discovery is of a composition of matter, the applicant, if required by the commissioner, shall furnish specimens of ingredients and of the composition, sufficient in quantity for the purpose of experiment.

SEC. 29. *And be it further enacted*, That in all cases which admit of representation by model, the applicant, if required by the commissioner, shall furnish one of convenient size to exhibit advantageously the several parts of his invention or discovery.

SEC. 30. *And be it further enacted*, That the applicant shall make oath or affirmation that he does verily believe himself to be the original and first inventor or discoverer of the art, machine, manufacture, composition or improvement for which he solicits a patent: that he does not know and does not believe that the same was ever before known or used; and shall state of what country he is a citizen. And said oath or affirmation may be made before any person in the United States authorized by law to administer oaths; or when the applicant resides in a foreign country, before any minister, *charge d'affaires*, consul, or commercial agent, holding commission under the government of the United States, or before any notary public of the foreign country in which the applicant may be.

SEC. 31. *And be it further enacted*, That on the filing of any such application and the payment of the duty required by law, the commissioner shall cause an examination to be made of the alleged new invention or discovery; and if on such examination it shall appear that the claimant is justly entitled to a patent under the law, and that the same is sufficiently useful and important, the commissioner shall issue a patent therefor.

SEC. 32. *And be it further enacted*, That all applications for patents shall be completed and prepared for examination within two years after the filing of the petition, and in default thereof, or upon failure of the applicant to prosecute the same within two years after any action therein, of which notice shall have been given to the applicant, they shall be regarded as abandoned by the parties thereto, unless it be shown to the satisfaction of the Commissioner that such delay was unavoidable.

SEC. 33. *And be it further enacted*, That patents may be granted and issued or re-issued to the assignee of the inventor or discoverer, the assignment thereof being first entered on record in the Patent Office; but in such case the application for the patent shall be made, and the specification sworn to, by the inventor or discoverer; and also, if he be living, in case of an application for re-issue.

SEC. 34. *And be it further enacted*, That when any person, having made any new invention or discovery for which a patent might have been granted, dies before a patent is granted, the right of applying for and obtaining the patent shall devolve on his executor or administrator, in trust for the heirs at law of the deceased, in case he shall have died intestate; or if he shall have left a will, disposing of the same, then in trust for his devisees, in as full manner and on the same terms and conditions as the same might have been claimed or enjoyed by him in his lifetime; and when the application shall be made by such legal representatives, the oath or affirmation required to be made shall be so varied in form that it can be made by them.

SEC. 35. *And be it further enacted*, That any person who has an interest in an invention or discovery, whether as inventor, discoverer, or assignee, for which a patent was ordered to issue upon the payment of the final fee, but who has failed to make payment thereof within six months from the time at which it was passed and allowed, and notice thereof was sent to the applicant or his agent, shall have a right to make an application for a patent for such invention or discovery the same as in the case of an original application : *Provided*, That the second application be made within two years after the allowance of the original application. But no person shall be held responsible in damages for the manufacture or use of any article or thing for which a patent, as aforesaid, was ordered to issue, prior to the issue thereof: *And provided, further*, That when an application for a patent has been rejected or withdrawn, prior to the passage of this act, the applicant shall have six months from the date of such passage to renew his application, or to file a new one, and if he omit to do either, his application shall be held to have been abandoned. Upon the hearing of such renewed applications abandonment shall be considered as a question of fact.

Sec. 36. *And be it further enacted*, That every patent or any interest therein shall be assignable in law, by an instrument in writing ; and the patentee or his assigns or legal representatives may, in like manner, grant and convey an exclusive right under his patent to the whole or any specified part of the United States ; and said assignment, grant, or conveyance. shall be void as against any subsequent purchaser or mortgagee for a valuable consideration, without notice, unless it is recorded in the Patent Office within three months from the date thereof.

Sec. 37. *And be it further enacted*, That every person who may have purchased of the inventor, or with his knowledge and consent may have constructed any newly invented or discovered machine, or other patentable article, prior to the application by the inventor or discoverer for a patent, or sold or used one so constructed, shall have the right to use, and vend to others to be used, the specific thing so made or purchased, without liability therefor.

Sec. 38. *And be it further enacted*, That it shall be the duty of all patentees, and their assigns and legal representatives, and of all persons making or vending any patented article for or under them, to give sufficient notice to the public that the same is patented, either by fixing thereon the word "patented,"

together with the day and year the patent was granted: or
when, from the character of the article, this can not be done,
by fixing to it or to the package wherein one or more of them
is inclosed, a label containing the like notice; and in any suit
for infringement, by the party failing so to mark, no damages
shall be recovered by the plaintiff, except on proof that the
defendant was duly notified of the infringement, and continued,
after such notice, to make, use, or vend the article so patented.

SEC. 39. *And be it further enacted*, That if any person shall, in
any manner, mark upon anything made, used or sold by him for
which he has not obtained a patent, the name or any imitation
of the name of any person who has obtained a patent therefor,
without the consent of such patentee, or his assigns or legal
representatives; or shall in any manner mark upon or affix to
any such patented article the word "patent" or "patentee,"
or the words "letters-patent," or any word of like import,
with intent to imitate or counterfeit the mark or device of the
patentee, without having the license or consent of such patentee
or his assigns or legal representatives; or shall in any manner
mark upon or affix to any unpatented article the word "patent,"
or any word importing that the same is patented, for the purpose of deceiving the public, he shall be liable for every such
offence to a penalty of not less than one hundred dollars, with
costs; one moiety of said penalty to the person who shall sue
for the same, and the other to the use of the United States, to
be recovered by suit in any district court of the United States
within whose jurisdiction such offence may have been committed.

SEC. 40. *And be it further enacted*. That any citizen of the
United States, who shall have made any new invention or discovery, and shall desire further time to mature the same, may,
on payment of the duty required by law, file in the Patent
Office a caveat setting forth the design thereof, and of its dis-
tinguishing characteristics, and praying protection of his right
until he shall have matured his invention; and such caveat
shall be filed in the confidential archives of the office and preserved in secrecy, and shall be operative for the term of one

year from the filing thereof, and if application shall be made within the year by any other person for a patent with which such caveat would in any manner interfere, the Commissioner shall deposit the description, specifications, drawings, and model of such application in like manner in the confidential archives of the office, and give notice thereof, by mail, to the person filing the caveat, who, if he would avail himself of his caveat shall file his description, specifications, drawings, and model within three months from the time of placing said notice in the post office in Washington, with the usual time required for transmitting it to the caveator added thereto, which time shall be indorsed on the notice   And an alien shall have the privilege herein granted, if he shall have resided in the United States one year next preceding the filing of his caveat, and made oath of his intention to become a citizen.

SEC. 41. *And be it further enacted,* That whenever, on examination, any claim for a patent is rejected for any reason whatever, the Commissioner shall notify the applicant thereof, giving him briefly the reasons for such rejection, together with such information and references as may be useful in judging of the propriety of renewing his application or of altering his specification; and if, after receiving such notice, the applicant shall persist in his claim for a patent, with or without altering his specifications, the Commissioner shall order a reëxaminion of the case.

SEC. 42. *And be it further enacted,* That whenever an application is made for a patent which, in the opinion of the Commissioner, would interfere with any pending application, or with any unexpired patent, he shall give notice thereof to the applicants, or applicant and patentee, as the case may be, and shall direct the primary examiner to proceed to determine the question of priority of invention.   And the Commissioner may issue a patent to the party who shall be adjudged the prior inventor, unless the adverse party shall appeal from the decision of the primary examiner, or of the board of examiners-in-chief, as the case may be, within such time, not less than twenty days, as the Commissioner shall prescribe.

Sec. 43. *And be it further enacted*, That the Commissioner may establish rules for taking affidavits and depositions required in cases pending in the Patent Office, and such affidavits and depositions may be taken before any officer authorized by law to take depositions to be used in the courts of the United States, or of the State where the officer resides.

Sec. 44. *And be it further enacted*, That the clerk of any court of the United States, for any district or territory wherein testimony is to be taken for use in any contested case pending in the Patent Office, shall, upon the application of any party thereto, or his agent or attorney, issue subpœna for any witness residing or being within said district or territory, commanding him to appear and testify before any officer in said district or territory authorized to take depositions and affidavits, at any time and place in the subpœna stated; and if any witness, after being duly served with such subpœna, shall neglect or refuse to appear, or after appearing shall refuse to testify, the judge of the court whose clerk issued the subpœna may, on proof of such neglect or refusal, enforce obedience to the process, or punish the disobedience as in other like cases.

Sec. 45. *And be it further enacted*, That every witness duly subpœnaed and in attendance shall be allowed the same fees as are allowed to witnesses attending the courts of the United States, but no witness shall be required to attend at any place more than forty miles from the place where the subpœna is served upon him, nor be deemed guilty of contempt for disobeying such subpœna, unless his fees and travelling expenses in going to, returning from, and one day's attendance at the place of examination, are paid or tendered him at the time of the service of the subpœna; nor for refusing to disclose any secret invention or discovery made or owned by himself.

Sec. 46. *And be it further enacted*, That every applicant for a patent or the re-issue of a patent, any of the claims of which have been twice rejected, and every party to an interference, may appeal from the decision of the primary examiner, or of the examiner in charge of interference, in such case to the board of examiners-in-chief, having once paid the fee for such appeal provided by law

SEC. 47. *And be it further enacted*, That if such party is dissatisfied with the decision of the examiners-in-chief, he may, on payment of the duty required by law, appeal to the Commissioner in person.

SEC. 48. *And be it further enacted*, That if such party, except a party to an interference, is dissatisfied with the decision of the Commissioner, he may appeal to the Supreme Court of the District of Columbia, sitting in banc.

SEC. 49. *And be it further enacted*, That when an appeal is taken to the Supreme Court of the District of Columbia the appellant shall give notice thereof to the Commissioner, and file in the Patent Office, within such time as the Commissioner shall appoint, his reasons of appeal, specifically set forth in writing.

SEC. 50. *And be it further enacted*, That it shall be the duty of said court, on petition, to hear and determine such appeal, and to revise the decision appealed from in a summary way, on the evidence produced before the Commissioner, at such early and convenient time as the court may appoint, notifying the Commissioner of the time and place of hearing, and the revision shall be confined to the points set forth in the reason of appeal. And after hearing the case, the court shall return to the Commissioner a certificate of its proceedings and decision, which shall be entered of record in the Patent Office, and govern the further proceedings in the case. But no opinion or decision of the court in any such case shall preclude any person interested from the right to contest the validity of such patent in any court wherein the same may be called in question.

SEC. 51 *And be it further enacted*, That on receiving notice of the time and place of hearing such appeal, the Commissioner shall notify all parties who appear to be interested therein, in such manner as the court may prescribe. The party appealing shall lay before the court certified copies of all the original papers and evidence in the case, and the Commissioner shall furnish it with the grounds of his decision, fully set forth in writing, touching all the points involved by the reasons of appeal. And at the request of any party interested, or of the court, the Commissioner and the examiners may be examined under oath in explanation of the principles of the machine or other thing for which a patent is demanded.

SEC. 52. *And be it further enacted,* That whenever a patent on application is refused, for any reason whatever, either by the Commissioner or by the Supreme Court of the District of Columbia upon appeal from the Commissioner, the applicant may have remedy by bill in equity ; and the court having cognizance thereof, on notice to adverse parties and other due proceedings had, may adjudge that such applicant is entitled, according to law, to receive a patent for his invention, as specified in his claim, or for any part thereof, as the facts in the case may appear. And such adjudication, if it be in favor of the right of the applicant, shall authorize the Commissioner to issue such patent, on the applicant filing in the Patent Office a copy of the adjudication, and otherwise complying with the requisitions of law. And in all cases where there is no opposing party a copy of the bill shall be served on the commissioner, and all the expenses of the proceeding shall be paid by the applicant, whether the final decision is in his favor or not.

SEC. 53. *And be it further enacted,* That whenever any patent is inoperative or invalid, by reason of a defective or insufficient specification, or by reason of the patentee claiming as his own invention or discovery more than he had a right to claim as new, if the error has arisen by inadvertence, accident or mistake, and without any fraudulent or deceptive intention, the Commissioner shall, on the surrender of such patent and the payment of the duty required by law, cause a new patent for the same invention, and in accordance with the corrected specification, to be issued to the patentee, or, in the case of his death or assignment of the whole or any undivided part of the original patent, to his executors, administrators, or assigns, for the unexpired part of the term of the original patent, the surrender of which shall take effect upon the issue of the amended patent; and the Commissioner may, in his discretion, cause several patents to be issued for distinct and separate parts of the thing patented, upon demand of the applicant, and upon payment of the required fee for a re-issue for each of such re-issued letters-patent. And the specifications and claim in every such case shall be subject to revision and restriction in the same manner as original applications are. And the patent so re-issued, together with the corrected specification, shall have the effect and operation in law, on the trial of all actions for causes thereafter arising, as though the same had been originally filed in such corrected forms; but no new matter shall be introduced into the specification, nor in case of a machine patent shall the model or drawings be amended, except each by the other; but when there is neither model nor drawing, amendments may be made upon proof satisfactory to the Commissioner that such new matter or amendment was a part of the original invention, and was omitted from the specification by inadvertence, accident, or mistake, as aforesaid.

SEC. 54. *And be it further enacted*, That whenever, through inadvertence, accident, or mistake, and without any fraudulent or deceptive intention, a patentee has claimed more than that of which he was the original or first inventor or discoverer, his patent shall be valid for all that part which is truly and justly his own, provided the same is a material or substantial part of the thing patented; and any such patentee, his heirs, or assigns, either of the whole or any sectional interest therein, may, on payment of the duty required by law, make disclaimer of such parts of the thing patented as he shall not choose to claim or to hold by virtue of the patent or assignment, stating therein the extent of his interest in such patent; said disclaimer shall be in writing, attested by one or more witnesses, and recorded in the Patent Office, and it shall thereafter be considered as part of the original specification to the extent of the interest possessed by the claimant and by those claiming under him after the record thereof. But no such disclaimer shall affect any action pending at the time of its being filed, except so far as may relate to the question of unreasonable neglect or delay in filing it.

SEC. 55. *And be it further enacted*, That all actions, suits, controversies, and cases arising under the patent laws of the

United States shall be originally cognizable, as well in equity as at law, by the circuit courts of the United States, or in any district court having the powers and jurisdiction of a circuit court, or by the Supreme Court of the District of Columbia, or of any territory ; and the court shall have power, upon bill in equity filed by any party aggrieved, to grant injunctions according to the course and principles of courts of equity, to prevent the violation of any right secured by patent, on such terms as the court may deem reasonable ; and upon a decree being rendered in any such case for an infringement, the claimant shall be entitled to recover, in addition to the profits to be accounted for by the defendant, the damages the complainant has sustained thereby, and the court shall assess the same or cause the same to be assessed under its direction, and the court shall have the same powers to increase the same in its discretion that are given by this act to increase the damages found by verdicts in actions upon the case ; but all actions shall be brought during the term for which the letters-patent shall be granted or extended, or within six years after the expiration thereof.

Sec. 56. *And be it further enacted,* That a writ of error or appeal to the Supreme Court of the United States shall lie from all judgments and decrees of any circuit court, or of any district court exercising the jurisdiction of a circuit court, or of the Supreme Court of the District of Columbia, or of any territory, in any action, suit, controversy, or case, at law or in equity, touching patent rights, in the same manner and under the same circumstances as in other judgments and decrees of such circuit courts, without regard to the sum of value in controversy.

Sec. 57. *And be it further enacted,* That written or printed copies of any records, books, papers, or drawings belonging to the Patent Office, and of letters-patent under the signature of the Commissioner or Acting Commissioner, with the seal of office affixed, shall be competent evidence in all cases wherein the originals could be evidence, and any person making application therefor, and paying the fee required by law, shall have certified copies thereof. And copies of the specifications and drawings of foreign letters-patent, certified in like manner, shall be *prima facie* evidence of the fact of the granting of such foreign letters-patent, and the date and contents thereof.

SEC. 58. *And be it further enacted*, That whenener there shall be interfering patents, any person interested in any of such interfering patents, or in the working of the invention claimed under either of such patents, may have relief against the interfering patentee, and all parties interested under him by suit in equity against the owners of the interfering patent ; and the court having cognizance thereof, as herein before provided, or notice to adverse parties, and other due proceedings had according to the course of equity, may adjudge and declare either of the patents void in whole or in part, or inoperative, or invalid in any particular part of the United States, according to the interest of the parties in the patent or the invention patented. But no such judgment or adjudication shall effect the rights of any person except the parties to the suit and those deriving title under them subsequent to the rendition of such judgment.

SEC. 59. *And be it further enacted*, That damages for the infringement of any patent may be recovered by action on the case in any circuit court of the United States, or district court exercising the jurisdiction of a circuit court, or in the Supreme Court of the District of Columbia, or of any territory, in the name of the party interested, either as patentee, assignee, or grantee. And whenever in any such action a verdict shall be rendered for the plaintiff, the court may enter judgment thereon for any sum above the amount found by the verdict as the actual damages sustained, according to the circumstances of the case, not exceeding three times the amount of such verdict, together with the costs.

SEC. 60. *And be it further enacted*, That whenever, through inadvertence, accident, or mistake, and without any willful default or intent to defraud or mislead the public, a patentee shall have (in his specification) claimed to be the original and first inventor or discoverer of any material or substantial part of the thing patented, of which he was not the original and first inventor or discoverer aforesaid. every such patentee, his executors, administrators, and assigns, whether of the whole or any sectional interest in the patent, may maintain a suit at at law or in equity for the infringement of any part thereof which was *bona fide* his own, provided it shall be a material and substantial part of the thing patented, and be definitely distinguishable from the parts so claimed, without right as aforesaid, notwithstanding the specifications may embrace more than that of which the patentee was the original or first inventor or discoverer. But in every such case in which a judgment or decree shall be rendered for the plaintiff, no costs shall be recovered unless the proper disclaimer has been entered at the Patent Office before the commencement of the suit ; nor shall he be entitled to the benefits of this section if he shall have unreasonably neglected or delayed to enter said disclaimer.

SEC. 61 *And be it further enacted,* That in any action for infringement the defendant may plead the general issue, and, having given notice in writing to the plaintiff or his attorney, thirty days before. may prove on trial any one or more of the following special matters :

First. That for the purpose of deceiving the public the description and specification filed by the patentee in the Patent Office was made to contain less than the whole truth relative to his invention or discovery. or more than is necessary to produce the desired effect : or,

Second. That he had surreptitiously or unjustly obtained the patent for that which was in fact invented by another, who was using reasonable diligence in adapting and perfecting the same ; or,

Third. That it has been patented or described in some printed publication prior to his supposed invention or discovery thereof: or,

Fourth. That he was not the original and first inventor or discoverer of any material and substantial part of the thing patented ; or,

Fifth. That it had been in public use or on sale in this country for more than two years before his application for a patent. or had been abandoned to the public.

And in notice as to proof of previous invention, knowledge, or use of the thing patented, the defendant shall state the names of patentees and the dates of their patents, and when granted, and the names and residences of the persons alleged to have invented or to have had the prior knowledge of the thing patented, and where and by whom it had been used ; and if any one or more of the special matters alleged shall be found for the defendant, judgment shall be rendered for him with costs. And the like defenses may be pleaded in any suit in equity for relief against an alleged infringement ; and proofs of the same may be given upon like notice in the answer of the defendant, and with the like effect.

SEC. 62. *And be it further enacted,* That whenever it shall appear that the patentee, at the time of making his application for the patent, believed himself to be the original and first inventor or discoverer of the thing patented, the same shall not be held to be void on account of the invention or discovery, or any part thereof, having been known or used in a foreign country, before his invention or discovery thereof, if it had not been patented, or described in a printed publication.

SEC. 63. *And be it further enacted,* That where the patentee of an invention or discovery, the patent for which was granted prior to the second day of March, eighteen hundred and sixty-one, shall desire an extension of his patent beyond the original term of its limitation, he shall make application therefor, in writing, to the commissioner, setting forth the reason why such extension should be granted ; and he shall also furnish a written statement under oath of the ascertained value of the invention or discovery, and of his receipts and expenditures on account thereof, sufficiently in detail to exhibit a true and faithful account of the loss and profit in any manner accruing to him by reason of said invention or discovery. And said application shall be filed not more than six months nor less than ninety days before the expiration of the original term of the patent, and no extension shall be granted after the expiration of said original term.

SEC. 64. *And be it further enacted,* That upon the receipt of such application, and the payment of the duty required by law, the commissioner shall cause to be published in one newspaper in the city of Washington, and in such other papers published in the section of the country most interested adversely to the extension of the patent as he may deem proper, for at least sixty days prior to the day set for hearing the case, a notice of such application, and of the time and place when and where the same will be considered, that any person may appear and show cause why the extension should not be granted.

SEC. 65. *And be it further enacted,* That on the publication of such notice, the commissioner shall refer the case to the principal examiner having charge of the class of inventions to which it belongs, who shall make to said commissioner a full report of the case, and particularly whether the invention or discovery was new and patentable when the original patent was granted.

SEC. 66 *And be it further enacted,* That the commissioner shall, at the time and place designated in the published notice, hear and decide upon 'the evidence produced, both for and against the extension ; and if it shall appear to his satisfaction that the patentee, without neglect or fault on his part, has failed to obtain from the use and sale of his invention or discovery a reasonable remuneration for the time, ingenuity, and expense bestowed upon it, and the introduction of it into use, and that it is just and proper, having due regard to the public interest, that the term of the patent should be extended, the said Commissioner shall make a certificate thereon, renewing and extending the said patent for the term of seven years from the expiration of the first term, which certificate shall be recorded in the Patent Office, and thereupon the said patent shall have the same effect in law as though it had been originally granted for twenty-one years.

SEC. 67. *And be it further enacted,* That the benefit of the extension of a patent shall extend to the assignees and grantees of the right to use the thing patented to the extent of their interest therein.

SEC. 68. *And be it further enacted,* That the following shall be the rates for patent fees:

On filing each original application for a patent, fifteen dollars.

On issuing each original patent, twenty dollars.

On filing each caveat, ten dollars.

On every application for the re-issue of a patent, thirty dollars.

On filing each disclaimer, ten dollars.

On every application for the extension of a patent, fifty dollars.

On the granting of every extension of a patent, fifty dollars.

On an appeal for the first time from the primary examiners to the examiners-in-chief, ten dollars.

On every appeal from the examiners-in-chief to the Commissioner, twenty dollars.

For certified copies of patents and other papers, ten cents per hundred words.

For recording every assignment, agreement, power of attorney, or other paper, of three hundred words or under, one dollar; of over three hundred and under one thousand words, two dollars; of over one thousand words, three dollars.

For copies of drawings, the reasonable cost of making them.

SEC. 69. *And be it further enacted*, That patent fees may be paid to the Commissioner or to the treasurer, or any of the assistant treasurers of the United States, or to any of the designated depositaries, national banks, or receivers of public money designated by the Secretary of the Treasury for that purpose, who shall give the depositor a receipt or certificate of deposit therefor. And all money received at the Patent Office, for any purpose, or from any source whatever, shall be paid into the treasury as received, without any deduction whatever; and all disbursements for said office shall be made by the disbursing clerk of the Interior Department.

SEC. 70. *And be it further enacted*, That the treasurer of the United States is authorized to pay back any sum or sums of money to any person who shall have paid the same into the treasury, or to any receiver or depositary, to the credit of the treasurer, as for fees accruing at the Patent Office through mistake, certificate thereof being made to said treasurer by the Commissioner of Patents.

SEC. 71. *And be it further enacted*, That any person who, by his own industry, genius, efforts, and expense, has invented or produced any new and original design for a manufacture, bust, statue, alto-relievo, or bas-relief; any new and original design for the printing of woolen, silk, cotton, or other fabrics; any new and original impression, ornament, pattern, print, or picture, to be printed, painted, cast, or otherwise placed on or worked into any article of manufacture; or any new, useful and original shape or configuration of any article of manufacture, the same not having been known or used by others before his invention or production thereof, or patented or described in any printed publication, may, upon payment of the duty required by law, and other due proceedings had the same as in cases of inventions or discoveries, obtain a patent therefor.

SEC. 72. *And be it further enacted*, That the Commissioner may dispense with models of designs when the design can be sufficiently represented by drawings or photographs.

SEC. 73. *And be it further enacted*, That patents for designs may be granted for the term of three years and six months, or for seven years, or for fourteen years, as the applicant may, in his application, elect.

SEC. 74. *And be it further enacted*, That patentees of designs issued prior to March two, eighteen hundred and sixty-one, shall be entitled to extension of their respective patents for the term of seven years, in the same manner and under the same restrictions as are provided for the extension of patents for inventions or discoveries, issued prior to the second day of March, eighteen hundred and sixty-one.

SEC. 75. *And be it further enacted*, That the following shall be the rates of fees in design cases:

For three years and six months, ten dollars.

For seven years, fifteen dollars.

For fourteen years, thirty dollars.

For all other cases in which fees are required, the same rates as in cases of inventions or discoveries.

SEC. 76. *And be it further enacted*, That all the regulations and provisions which apply to the obtaining or protection of patents for inventions or discoveries, not inconsistent with the provisions of this act, shall apply to patents for designs.

SEC. 77. *And be it further enacted,* That any person or firm domiciled in the United States, and any corporation created by the authority of the United States, or of any State or territory thereof, and any person, firm, or corporation resident of or located in any foreign country which by treaty or convention affords similar privileges to citizens of the United States, and who are entitled to the exclusive use of any lawful trade-mark, or who intend to adopt and use any trade-mark for exclusive use within the United States, may obtain protection for such lawful trade-mark by complying with the following requirements, to wit :

First. By causing to be recorded in the Patent Office the names of the parties and their residences and place of business. who desire the protection of the trade-mark.

Second. The class of merchandise and the particular description of goods comprised in such class, by which the trade-mark has been or is intended to be appropriated.

Third. A description of the trade-mark itself, with *fac similes* thereof, and the mode in which it has been or is intended to be applied or used.

Fourth. The length of time, if any, during which the trade-mark has been used.

Fifth. The payment of a fee of twenty-five dollars, in the same manner and for the same purpose as the fee required for patents.

Sixth. The compliance with such regulations as may be prescribed by the Commissioner of Patents.

Seventh. The filing of a declaration, under the oath of the person, or of some member of the firm, or officer of the corporation, to the effect that the party claiming protection for the trade-mark has a right to the use of the same, and that no other person, firm or corporation has the right to such use, either in the identical form, or having such near resemblance thereto as might be calculated to deceive, and that the description and *fac-similes* presented for record are true copies of the trade-mark sought to be protected.

SEC. 78. *And be it further enacted,* That such trade-mark shall

remain in force for thirty years from the date of such registration, except in cases where such trade-mark is claimed for and
applied to articles not manufactured in this country and in
which it receives protection under the laws of any foreign
country for a shorter period, in which case it shall cease to
have any force in this country by virtue of this act at the same
time that it becomes of no effect elsewhere; and during the
period that it remains in force it shall entitle the person, firm
or corporation registering the same to the exclusive use thereof
so far as regards the description of goods to which it is appropriated in the statement filed under oath as aforesaid, and no
other person shall lawfully use the same trade-mark, or substantially the same, or so nearly resembling it as to be calculated
to deceive, upon substantially the same description of goods :
Provided, That six months prior to the expiration of said term
of thirty years, application may be made for a renewal of such
registration, under regulations to be prescribed by the Commissioner of Patents, and the fee for such renewal shall be the
same as for the original registration; certificate of such renewal
shall be issued in the same manner as for the original registration, and such trade-mark shall remain in force for a further
term of thirty years: *And provided further,* That nothing in this
section shall be construed by any court as abridging or in any
manner affecting unfavorably the claim of any person, firm,
corporation or company, to any trade-mark after the expiration
of the term for which such trade-mark was registered.

SEC. 79. *And be it further enacted,* That any person or corporation who shall reproduce, counterfeit, copy, or imitate any
such recorded trade-mark, and affix the same to goods of substantially the same descriptive properties and qualities as those
referred to in the registration, shall be liable to an action in
the case for damages for such wrongful use of said trade-mark,
at the suit of the owner thereof, in any court of competent
jurisdiction in the United States, and the party aggrieved shall
also have his remedy according to the course of equity to enjoin
the wrongful use of his trade-mark and to recover compensation
therefor in any court having jurisdiction over the person guilty

of such wrongful use. The Commissioner of Patents shall not receive and record any proposed trade-mark which is not and cannot become a lawful trade-mark, or which is merely the name of a person, firm, or corporation only, unaccompanied by a mark sufficient to distinguish it from the same name when used by other persons, or which is identical with the trade-mark appropriate to the same class of merchandise and belonging to a different owner, and already registered or received for registration, or which so nearly resembles such last-mentioned trade-mark as to be likely to deceive the public: *Provided*, That this section shall not prevent the registry of any lawful trade-mark rightfully used at the passage of this act.

SEC. 80. *And be it further enacted*, That the time of the receipt of any trade-mark at the Patent Office for registration shall be noted and recorded, and copies of the trade-mark and of the date of the receipt thereof, and of the statement filed therewith, under the seal of the Patent Office, certified by the Commissioner, shall be evidence in any suit in which such trade-mark shall be brought in controversy.

SEC. 81. *And be it further enacted*, That the Commissioner of Patents is authorized to make rules, regulations, and prescribe forms for the transfer of the right to the use of such trade-marks, conforming as nearly as practicable to the requirements of law respecting the transfer and transmission of copy-rights.

SEC. 82. *And be it further enacted*, That any person who shall procure the registry of any trade-mark, or of himself as the owner thereof, or an entry respecting a trade-mark in the Patent office under this act, by making any false or fraudulent representations or declarations, verbally or in writing, or by any fraudulent means, shall be liable to pay damages in consequence of any such registry or entry to the person injured thereby, to be recovered in an action on the case before any court of competent jurisdiction within the United States.

SEC. 83. *And be it further enacted*, That nothing in this act shall prevent, lessen, impeach, or avoid any remedy at law or in equity, which any party aggrieved by any wrongful use of any trade-mark might have had if this act had not been passed.

SEC. 84. *And be it further enacted*, That no action shall be maintained under the provisions of this act by any person claiming the exclusive right to any trade-mark which is used or claimed in any unlawful business, or upon any article which is injurious in itself, or upon any trade-mark which has been fraudulently obtained, or which has been formed and used with the design of deceiving the public in the purchase or use of any article of merchandise.

SEC. 111. *And be it further enacted*, That the acts and parts of acts set forth in the schedule of acts cited, hereto annexed, are hereby repealed, without reviving any acts or parts of acts repealed by any of said acts, or by any clause or provision therein: *Provided, however*, That the repeal hereby enacted shall not affect, impair, or take away any right existing under any of said laws; but all actions and causes of action, both in law and in equity, which have arisen under any of said laws, may be commenced and prosecuted; and, if already commenced, may be prosecuted to final judgment and execution, in the same manner as though this act had not been passed, excepting that the remedial provisions of this act shall be applicable to all suits and proceedings hereafter commenced: *And provided also*, That all applications for patents pending at the time of the passage of this act, in cases where the duty has been paid, shall be proceeded with and acted on in the same manner as though filed after the passage thereof: *And provided further*, That all offenses which are defined and punishable under any of said acts, and all penalties and forfeitures created thereby, and incurred before this act takes effect, may be prosecuted, sued for, and recovered, and such offenses punished according to the provisions of said acts, which are continued in force for such purpose.

Schedule of statutes cited and repealed, as printed in the Statutes at Large, including such portions only of the appropriation bills referred to as are applicable to the Patent Office.

Act of July 4, 1836, chapter 357, volume 5, page 117.
March 3, 1837, chapter 45, volume 5, page 191.

Act of March 3, 1839, chapter 88, volume 5, page 353.

August 29, 1842, chapter 263, volume 2, page 543.

August 6, 1846, chapter 90, volume 9, page 59.

May 27, 1848, chapter 47, volume 9, page 231.

March 3, 1849, chapter 108, volume 9, page 395.

March 3, 1851, chapter 42, volume 9, page 617.

August 30, 1852, chapter 107, volume 10, page 75.

August 31, 1852, chapter 108, volume 10, page 209.

April 22, 1854, chapter 52, volume 10, page 276.

March 3, 1855, chapter 175, volume 10, page 643.

August 18,1856, chapter 129, volume 11, page 81.

March 2, 1861, chapter 88, volume 12, page 246.

March 3, 1863, chapter 102, volume 12, page 796.

June 25, 1864, chapter 159, volume 13, page 194.

March 3, 1865, chapter 112, volume 13 page 533.

June 27, 1866, chapter 143, volume 14, page 76.

March 29, 1867, chapter 17, volume 15, page 10.

July 20, 1868, chapter 177, volume 15, page 119.

July 23, 1868, chapter 227, volume 15, page 168.

March 3, 1868, chapter 121, volume 15, page 293.

ABSTRACT

OF THE

POPULATION OF THE UNITED STATES OF AMERICA.

CENSUS OF 1870.

MASSACHUSETTS.—Area, 7,800 square miles.

Barnst'ble... 32774	Essex.........200843	Middlesex...274353	Suffolk 270802
Berkshire... 64827	Franklin..... 32635	Nantucket.. 4134	Worcester...192716
Bristol.......102886	Hampden... 78409	Norfolk...... 89443	
Dukes 3787	Hampshire 44388	Plymouth... 65365	Total... 1457351

NEW HAMPSHIRE.—Area, 9,280 square miles.

Belknap 17681	Coos............ 14932	Merrimac... 42151	Sullivan...... 18058
Carroll. 17332	Grafton...... 39103	Rock'h'm ... 47298	
Cheshire..... 27265	Hillsboro'.. 64328	Strafford..... 30242	Total 317710

OHIO—Area, 39,964 square miles.

Adams 21140	Fayette...... 17181	Loraine...... 30438	Richland ... 31970
Allen 23547	Franklin..... 63524	Lucas......... 44193	Ross............ 37090
Ashland 21922	Fulton 17796	Madison 15636	Sandusky... 25566
Ashtabula... 32427	Gallia......... 25421	Mahoning ... 30634	Scioto........ 28385
Athens....... 23889	Geauga....... 13084	Marion....... 16291	Seneca....... 30846
Auglaize..... 20043	Greene....... 29516	Medina 20082	Shelby 20754
Belmont..... 39913	Guernsey ... 23903	Meigs......... 31284	Stark......... 52708
Brown........ 30853	Hamilton....260617	Mercer....... 17268	Summit ... 34986
Butler 39953	Hancock..... 23803	Miami........ 32747	Trumbull... 38354
Carroll 14501	Hardin....... 18615	Monroe..... 25813	Tuscara's ... 33836
Champa'n... 24210	Harrison 18640	Montgo'y.... 60409	Union......... 12793
Clark......... 32117	Henry........ 13928	Morgan...... 20247	Van Wert... 15709
Clermont ... 34308	Highland ... 29163	Morrow....... 18581	Vinton 15047
Clinton 21921	Hocking..... 17934	Muskin'm ... 45200	Warren...... 26709
Columbia... 38655	Holmes...... 18176	Noble 19956	Washington 39979
Coshocton... 23647	Huron........ 28525	Ottawa....... 13244	Wayne 35634
Crawford ... 24588	Jackson 21859	Paulding ... 8552	Williams.... 21028
Cuyahoga...133105	Jefferson..... 29191	Perry......... 18465	Wood...... 24671
Darke........ 30972	Knox......... 25405	Pickaway... 24274	Wyandotte.. 18563
Defiance 15722	Lake 15953	Pike........... 15540	
Delaware.... 25187	Lawrence... 39600	Portage...... 24194	Total ... 2652302
Erie 28206	Licking...... 37707	Preble...... ... 21833	
Fairfield 31184	Logan....... 23084	Putnam..... 17104	

OREGON—Area, 102,606 square miles.

Baker 2804	Curry 504	Linn 8717	Union.......... 2552
Benton........ 4584	Douglas 6066	Marion......... 9966	Wasco'......... 2488
Clackam's ... 5993	Grant.......... 2251	Multnomah ..11510	Washington.. 4261
Clatsop 1254	Jackson....... 4778	Polk............ 4710	Yamhill....... 4989
Columbia 863	Josephine..... 1204	Tillamook 408	
Coos............ 1644	Lane 6426	Umatilla...... 2916	Total 90878

CONNECTICUT—Area, 4,674 square miles.

Fairfield..... 95272	Litchfield... 48727	NewHaven.121257	Tolland...... 22000
Hartford....109006	Middlesex... 36099	NewLo'don. 66534	Windham... 38518

Total..... 537417

WISCONSIN—Area, 53,924 square miles.

Adams........ 6605	Douglas...... 1121	Manitowoc...33185	Sauk............28853
Ashland....... 221	Dunn.......... 9422	Marathon.... 5885	Shawano..... 3165
Barron........ 538	Eau Claire...10893	Marquette... 8018	Sheboygan...31759
Bayfield...... 344	Fond du Lac.46247	Milwaukee...89956	St. Croix.....11035
Brown........25199	Grant.........38564	Monroe.......16551	Tremp'leau...10731
Buffalo........11075	Green.........23675	Oconto........ 8321	Vernon......18645
Burnett...... 776	Green Lake..13165	Outagamie...18434	Walworth....25967
Calumet......12329	Iowa.........24532	Ozaukee.......15568	Washington.23930
Chippewa..... 8345	Jackson...... 7707	Pepin........... 4661	Waukesha...28324
Clark.......... 3450	Jefferson 34065	Pierce......... 9995	Waupaca......15559
Columbia 28820	Juneau.......12394	Polk............ 3496	Waushara....11287
Crawford....13080	Kenosha......13149	Portage.......10751	Winnebago...37323
Dane.........53404	Kewaunee...10130	Rocine......26740	Wood.......... 3912
Dodge........47012	La Crosse.....20255	Richland......15332	
Door.......... 4922	Lafayette....,22651	Rock..........39034	Total..... 1055501

MAINE.—Area, 31,766 square miles.

Androsco'n. 35390	Kennebec... 53202	Piscataq's... 14095	York........ 60195
Aroostook... 30210	Knox......... 31831	Sagadaho'... 18820	
Cumberl'd... 82103	Lincoln....... 25670	Somerset.... 34644	Total......628719
Franklin.... 18699	Oxford....... 33515	Waldo....... 34456	
Hancock.... 36522	Penobscot... 70668	Washin'gn.. 43526	

MICHIGAN.—Area, 56,423 square miles.

Alcona....... 696	Emmett...... 1211	Lenawee...., 45596	Ontonagon.. 2845
Allegan..... 32106	G'd Trav'rse 4443	Livingston.. 19336	Osceola...... 2073
Alpena...... 2756	Genesee..... 33900	Mackinaw... 1716	Oscoda........ 70
Antrim...... 1985	Gratiot....... 11810	Macomb..... 27616	Ottawa...... 26649
Barry........ 22202	Hillsdale ... 31684	Manistee.... 6074	Presque Isl. 355
Busque Isle 355	Houghton... 13879	Manitou..... 891	Saginaw..... 39097
Bay.......... 15900	Huron........ 9053	Marquette... 15033	Sanilac....... 14562
Benzie........ 2184	Ingham...... 25268	Mason........ 3264	Shiawassee. 20858
Berrian..... 35104	Ionia........ 27679	Mecosta..... 5643	St. Clair.... 36661
Branch....... 26226	Iosco......... 3163	Menominec. 1892	St. Joseph... 26276
Calhoun 36569	Isabella....... 4113	Midland...... 3285	Tuscola....... 13714
Cass 21094	Jackson...... 36050	Missaukee... 130	Van Buren. 28828
Charlevoix . 1724	Kalamazoo. 32054	Monroe...... 27483	Weshtenaw 41434
Cheboygan.. 2190	Kalkaskia... 424	Montcalm... 13629	Wayne......119041
Chippewa.... 1689	Keewenaw.. 4205	Muskegon... 14895	Wexford..... 650
Clare 366	Kent......... 50403	Newago...... 7291	
Clinton 22845	Lake 548	Oakland 40867	Total....1184653
Delta 2441	Lapeer....... 21355	Oceana...... 7222	
Eaton........ 25172	Leelanaw... 4816	Ogemaw..... 12	

IOWA.—Area, 50,914 square miles.

Adair.......... 3983	Decatur..... 11986	Jones....... 19775	Poweshick.. 15583
Adams....... 4453	Delaware.... 17434	Keokuk..... 19497	Ringgold.... 5684
Allamakee.. 16872	Des Moines. 27183	Kossnth...... 3360	Sac............ 1455
Appanoose.. 16480	Dickinson... 1351	Lee........... 37252	Scott......... 38559
Audubon.... 1212	Dubuque.... 37879	Linn......... 28818	Shelby...... 2540
Benton...... 22213	Emmett...... 1392	Louisa....... 13032	Sioux........ 577
Blackhawk. 21306	Fayette...... 16919	Lucas......... 10401	Story 11662
Boone........ 14569	Floyd......... 10770	Lyon.......... 221	Tama......... 16073
Bremer 12565	Franklin..... 4738	Madison...... 13811	Taylor....... 6990
Buchanan... 17042	Fremont..... 10976	Mohaska.... 22178	Union........ 5987

Buena Vista 1411	Greene 4635	Marion 24452	Van Buren 17695
Butler 9953	Grundy 6475	Marshall 16709	Wapello 22152
Calhoun 1602	Guthrie 7063	Mills 8678	Warren 17791
Carroll 2451	Hamilton 6051	Mitchel 9524	Washington 19223
Cass 5464	Hancock 1007	Mocona 3699	Wayne 11288
Cedar 19702	Hardin 13054	Monroe 12813	Webster 10550
Cerro Gordo 3415	Harrison 8921	Montgom'ry 5895	Winnebago 1572
Cherokee 1967	Henry 21484	Muscatine 21897	Win'shiek 23604
Chickasaw 10133	Howard 6281	O'Brien 715	Woodbury 6110
Clarke 8736	Humboldt 2575	Osceola(with Lyon)	Worth 2892
Clay 1523	Ida 226	Page 9977	Wright 2390
Clayton 27779	Iowa 16643	Palo Alto 1336	
Clinton 33994	Jackson 21462	Plymouth 1999	Total 1181359
Crawford 2612	Jasper 20787	Pocahontas 1448	
Dallas 12020	Jefferson 17861	Polk 27896	
Davis 15537	Johnson 24968	Pot'watme 16534	

ILLINOIS.—Area, 55,405 square miles.

Adams 56116	Ford 9103	Livingston 34392	Randolph 20859
Alexander 10519	Franklin 12668	Logan 23149	Richland 12803
Bond 12322	Fulton 38891	Macon 26481	Rock Island 29842
Boone 13007	Gallatin 11136	Macoupin 32771	Saline 12714
Brown 12212	Green 19665	Madison 44322	Sangamon 46384
Bureau 32430	Grundy 14974	Marion 20650	Schuyler 17419
Calhoun 6566	Hamilton 13014	Marshall 16959	Scott 10530
Carroll 16709	Hancock 35996	Mason 16250	Shelby 25529
Cass 9651	Hardin 5113	Massac 9581	St. Clair 51069
Champa'n 32738	Henderson 12600	McDon'gh 26563	Stark 10790
Christian 20362	Henry 35495	McHenry 23812	Stephenson 30078
Clark 18721	Iroquois 25789	McLean 53948	Tazewell 29850
Clay 15886	Jackson 19643	Menard 11756	Union 16555
Clinton 16280	Jo Davies 27831	Mercer 19270	Vermilion 30376
Coles 25285	Jasper 11234	Monroe 13009	Wabash 8841
Cook 350236	Jefferson 17914	Montgo'ry 25315	Warren 23070
Crawford 13897	Jersey 15054	Morgan 28501	Washington 17727
Cumber'd 12223	Johnson 11248	Moultrie 9997	Wayne 19758
De Kalb 23275	Kane 39068	Ogle 27539	White 16846
De Witt 14781	Kankakee 24394	Peoria 36601	Whiteside 27512
Douglas 13494	Kendall 12398	Perry 13723	Will 43020
Du Page 16761	Knox 39186	Piatt 10896	Williamson 17171
Edgar 21449	Lake 21033	Pike 30793	Winnebago 29372
Edwards 7593	LaSalle 61130	Pope 11441	Woodford 18980
Effingham 15609	Lawrence 12536	Pulaski 9127	
Fayette 19693	Lee 27252	Putnam 6295	Total 2529410

PENNSYLVANIA.—Area 46,000 square miles.

Adams 30315	Clinton 23213	Lancaster 121426	Potter 11424
Allegheny 262383	Columbia 28765	Lawrence 27298	Schuylkil 109325
Armstrong 43385	Crawford 63827	Lebanon 34117	Snyder 15606
Beaver 36132	Cumberland 43885	Lehigh 56792	Somerset 28233
Bedford 28636	Dauphin 60737	Luzerne 160951	Sullivan 6191
Berks 106739	Delaware 39541	Lycoming 47633	Susq'b'na 37530
Blair 38051	Elk 8315	McKean 8826	Tioga 35102
Bradford 53109	Erie 65977	Mercer 49981	Union 15568
Bucks 61997	Fayette 43284	Mifflin 17509	Venango 47522
Butler 36485	Forest 4183	Monroe 18389	Warren 23897
Cambria 36572	Franklin 45388	Montg'mry 81612	Washington 48481
Cameron 4273	Fulton 9361	Montour 15334	Wayne 33210
Carbon 28208	Greene 25883	Northamt'n 61403	Westm'rld 58699
Centre 33394	Huntingd'n 31252	Northum'ld 41440	Wyoming 14585
Chester 77824	Indiana 36123	Perry 25486	York 76217
Clarion 26542	Jefferson 21661	Philadel 673726	
Clearfield 25779	Juniata 17491	Pike 8414	Total 3511543

NEW JERSEY.—Area, 3,320 square miles.

Atlantic...... 14093	Essex..........143850	Monmouth.. 42821	Sussex........ 23242
Bergen......... 30124	Gloucester.. 21562	Morris........ 43137	Union......... 38966
Burlington . 57389	Hudson......128275	Ocean......... 13630	Warren....... 34338
Camden...... 46200	Hunterdon . 36965	Passaic...... 46423	
Cape May... 8199	Mercer....... 46386	Salem........ 23940	Total ... 903044
Cumberland 34658	Middlesex... 45034	Somerset 23512	

RHODE ISLAND.—Area, 1,306 square miles.

Bristol......... 9421	Newport..... 20050	Washington 20097	Total 217356
Kent.......... 18595	Providence 149193		

DELAWARE.—Area, 2,120 square miles.

Kent.......... 29804	Newcastle... 63515	Sussex........ 31696	Total......... 125015

MARYLAND.—Area, 11,124 square miles.

Alleghany... 38544	Cecil.......... 25888	Kent.......... 17256	Talbot........ 16157
Anne Arund 24513	Charles 15751	Montgomer. 20572	Washington 34714
Baltimore.. 330618	Dorchester.. 19598	P. George's. 29957	Wiconico.... 15844
Calvert 9856	Frederick... 47687	Qu. Anne's.. 15033	Worcester... 16472
Caroline...... 12121	Harford...... 23308	St. Mary's... 15089	
Carroll....... 26710	Howard...... 14167	Somerset 18200	Total...... 790095

VERMONT.—Area, 10,212 square miles.

Addison..... 23184	Essex......... 6811	Orange....... 23090	Windham.... 26036
Bennington 21325	Franklin 30291	Orleans....... 21035	Windsor 36064
Caledonia... 22217	Grand Isle.. 4082	Rutland...... 40671	
Chittenden. 36486	Lamoille..... 12448	Washington 26518	Total 330582

WEST VIRGINIA.—Area,   square miles.

Barbour...... 9200	Hancock..... 4363	Mineral...... 6349	Roane......... 7232
Berkeley 14900	Hardy 5518	Monongalia 13549	Taylor........ 9367
Boone........ 4553	Harrison 14200	Monroe....... 11124	Tucker........ 1907
Braxton 6181	Jackson...... 10300	Morgan....... 4315	Tyler.......... 7830
Brooke....... 5465	Jefferson 13219	Nicholas 4458	Upshur 8023
Cabell......... 6429	Kanawaha.. 22350	Ohio......... 28832	Wayne........ 7852
Calhoun...... 3000	Lewis......... 11375	Pendleton... 6455	Webster...... 1730
Clay........... 2196	Lincoln 5053	Pleasants.... 3012	Wetzel...... 8595
Doddridge... 7076	Logan 5124	Pocahontas. 4036	Wirt 4805
Fayette...... 6647	McDowell.... 1952	Preston....... 14554	Wood. 19010
Gilmer 4339	Marion....... 13200	Putnam...... 7794	Wyoming.... 3571
Grant........ 4468	Marshall..... 13300	Raleigh...... 3673	
Greenbrier., 13000	Mason 15958	Randolph.... 5563	Total 441094
Hampshire.. 7643	Mercer....... 7094	Ritchie...... 9055	

TENNESSEE.—Area, 45,600 square miles.

Anderson..... 8704	Franklin......14970	Lewis 1986	Scott 4054
Bedford.......24334	Gibson.........21700	Lincoln........28132	Sequatch'. ... 2335
Benton 8234	Giles...........32413	Macon.......... 6634	Sevier 11028
Bledsoe........ 4870	Granger.......12461	Madison 23526	Shelby 48000
Blount 14237	Greene 21668	Marion 6868	Smith 15994
Bradley......11652	Grundy........ 3583	Marshall......16207	Stewart 12029
Campbell 7445	Hamilton.....17241	Maury.........36285	Sullivan 13136
Cannon.......10502	Hancock...... 7148	McMinn......13969	Sumner........23711
Carroll.......19643	Hardeman ...17700	McNairy.....12727	Tipton.......14884
Carter........ 7909	Hardin.........11769	Meigs 4412	Union 7605
Cheatham.... 6678	Hawkins 15848	Monroe 12589	Van Buren... 2785
Claiborne 9321	Haywood 25095	Montgomery 24788	Warren.......12716
Cocke 12458	Henderson....14221	Morgan........ 2969	Washington..16318
Coffee 10237	Henry 20387	Obion 16684	Wayne 10209
Cumberland.. 3461	Hickman...... 9856	Overton 10988	Weakley......18200
Davidson 62906	Humphrey's. 9326	Perry........... 6903	White 9229

Decatur........ 6200	Jackson 13797	Polk............ 7369	Wilson 25914
De Kalb ...11425	Jefferson......19476	Putnam 8698	Williamson ..25354
Dickson........ 9340	Johnson...... 5852	Rhea............ 4854	
Dyer 13716	Knox 28994	Roane 15623	Total 1225937
Fayette 24300	Lauder'le 10839	Robertson ...16198	
Fentress....... 4717	Lawrence..... 7600	Rutherford...39844	

VIRGINIA.—Area, 61,352 square miles.

Accomack ...20409	Elizb. City... 8303	London.......20929	Richmond ... 6303
Albemarle ...27544	Essex.......... 9927	Louisa.......16332	Roanoke...... 6338
Alexandria ..16755	Fairfax 12952	Lunenb'g 10403	Rockbridge.. 5547
Alleghany ... 3674	Fauquier 19690	Madison 8670	Rock'gham...23668
Amelia 9878	Floyd............ 9824	Matthews..... 7000	Russell 11108
Amherst......14900	Fluvanna 9775	Mecklnb'g ...21318	Scott 13036
Appomat'x .. 8950	Franklin 20000	Middlesex ... 4981	Shenandoah .14936
Augusta 28763	Frederick ...16596	Montgomery 9385	Smyth........ 8898
Bath 3793	Giles 5659	Nansem'd.....11576	South'mpton12285
Bedford.......25327	Gloucester ...10211	Nelson........13898	Spottsylv'a...11728
Bland 4000	Goochland ...10312	New Kent.... 4378	Stafford........ 6420
Botetourt.....11329	Grayson....... 7586	Norfolk.......41580	Surry 5585
Brunswick ..13427	Greene 4634	North'pton... 8046	Sussex......... 7885
Buchanan ... 3777	Greencast..... 6362	North'land... 6863	Tazewell......10791
Buckingham 13369	Halifax........27828	Nottoway..... 9291	Warren........ 5716
Campbell 28384	Hanover......16456	Orange 10384	Warwick 1672
Caroline 15128	Henrico 66017	Page 8463	Washington..16816
Carroll 9147	Henry........12303	Patrick 10161	Westm'ld..... 7690
Chas. City ... 4977	Highland..... 4152	Pittsylv'a ...31354	Wise 4785
Charlotte 14513	Isle of Wight 8320	Powhatan ... 7667	Wythe........11611
Chesterfield..18470	James City... 4425	P. Edward...12004	York 7198
Clarke......... 6670	King &Queen 9709	P. George ... 7820	
Craig........... 2942	King George. 5742	P. William... 7505	Total 1211442
Culpepper ...12227	King Wm..... 7515	Pr'cess Anne 8273	
Cumberland.. 8142	Lancaster..... 5355	Pulaski....... 6538	
Dinwiddie ...30703	Lee..............13268	Rappah'nk... 8261	

NORTH CAROLINA.—Area, 50,704 square miles.

Almance......11874	Cumberland 17036	Lenoir 10436	Rockingb'm. 7869
Alexander ... 6868	Currituck..... 5131	Lillington .. ——	Rowan.........16811
Alleghany ... 3691	Dare 2778	Lincoln........ 9573	Rutherford...13121
Anson.........12428	Davidson 17256	Macon 6615	Sampson......16424
Ashe 1412	Davie........... 9620	Madison....... 8192	Stanly......... 8315
Beaufort......13054	Duplin 15542	Martin........ 9648	Stokes........ 6851
Bertie 12952	Edgecomb ...22971	McDowell..... 7592	Surry 11251
Bladen 12832	Forsyth........13050	Mecklinb'g...19181	Transylvania 3538
Brunswick... 7757	Franklin......14134	Mitchell 4705	Tyrrel 4173
Buncome 15419	Gaston 12602	Montgom'y... 7487	Union 12219
Burke 9777	Gates 7724	Moore 12040	Wake 35619
Cabarrus......11954	Granville 24831	Nash...........11077	Warren........17448
Caldwell...... 8476	Greene 8687	N. Hanover..27978	Washington.. 6516
Camden 5361	Guilford 21758	North'mpt'n.14749	Watauga 5287
Carteret 9010	Halifax........19000	Onslow 7569	Wayne........14000
Caswell.......16081	Hartnett 8895	Orange........17507	Wilkes.........12307
Catawba 10984	Haywood 7921	Pasquot'n'k.. 8131	Wilson.........12258
Chatham 15723	Henderson ... 7705	Perquimans.. 4336	Yadkin 10697
Cherokee 8080	Hertford 9273	Person.........11170	Yancey 5910
Chowan........ 6450	Hyde 6445	Pitt.............17276	
Clay............ ——	Iredell.........16931	Polk............ 4319	Totol 1016954
Cleveland.....12696	Jackson 6683	Randolph.....17555	
Columbus. ... 8474	Johnston 12713	Richmond...12882	
Craven 20516	Jones 5002	Robeson.......11069	

MISSISSIPPI.—Area, 47,156 square miles.

Adams...... ..	20000	Greene.......	2035	Lee...........	15954	Simpson	5718
Alcorn.......	10431	Grenada.....	10571	Lowndes.....	30504	Smith.........	7126
Amite.......	12000	Hancock.....	4240	Madison.....	23300	Sunflower...	5015
Attala.......	14774	Harrison...	5794	Marion.......	4211	Tallahat'e...	7852
Bolivar......	9732	Hinds........	30478	Marshall.....	29423	Tippah.......	20727
Calhoun.....	10561	Holmes......	19371	Monroe......	22632	Tishan'go...	24100
Carroll.......	2;671	Issaquena...	6888	Neshoba.....	8300	Tunica.......	5358
Chickasaw..	19891	Itawamba...	7812	Newton......	9774	Warren.....	26763
Choctaw.....	16000	Jackson......	4363	Noxubu......	20905	Washingt'n	14569
Claiborne ...	13385	Jasper.......	10884	Oktibbeha...	13000	Wayne......	4206
Clark.........	7505	Jefferson....	13848	Panolu.......	13800	Wilkinson...	15900
Coahoma....	7144	Jones........	3313	Perry........	2696	Winston.....	8984
Copiah......,	15400	Kemper......	11700	Pike.........	11363	Yalobusha..	13254
Covington...	4753	Lafayette ...	16100	Pontotoc.....	22100	Yazoo........	22300
Davis.........	—	Landerd'e...	13462	Prentiss......	9347		
De Soto......	23300	Lawrence....	6720	Rankin	12977	Total	842056
Franklin....	7198	Lake.........	9300	Scott	7848		

ARKANSAS.—Area, 52,198 square miles.

Arkansas ...	8268	Desha........	6125	Madison......	7927	St. Francis...	6714
Ashley.......	8042	Drew........ ..	9960	Marion.......	3979	Saline........	3911
Benton	13831	Franklin.....	9127	Mississippi..	3683	Scott	7483
Boone	—	Fulton..	4843	Monroe......	8336	Searcy.......	5614
Bradley.....	8646	Grant........	3943	Montg'me'y	2984	Sebastian ...	12940
Calhoun	3853	Greene.......	7573	Newton......	3364	Sevier	4492
Carroll	5780	Hempstead.	13768	Ouachita.....	12075	Sharpe.......	—
Chicot.......	7214	Hot Springs	5877	Perry........	2685	Union	10571
Clark........	11953	Independen	14566	Phillips	14800	Van Buren.	5107
Columbia...	11397	Izard........ .	6806	Pike..........	3788	Washington	17266
Conway.....	8112	Jackson	7268	Poinsett......	1720	White	10346
Crawford....	8957	Jefferson....	15733	Polk	3376	Woodruff....	6891
Crittenden..	4900	Johnson.....	9152	Pope........	8409	Yell........ ..	8048
Craighead...	4577	Lafayette ...	9139	Prairie.......	8800		
Cross	3915	Lawrence....	5981	Pulaski	32066	Total	473174
Dallas........	5707	Little River	3236	Randolph...	6200		

CALIFORNIA.—Area, 188,981 square miles.

Alpine.......	685	Klamath....	1674	Plumas......	4490	Siskiyou	6859
Amador......	9582	Lake.........	2969	Sacramento	26831	Solano......	16871
Alameda	24237	Lassen.......	1327	SanBer'd'no	3988	Sonoma......	19821
Butte........	11403	Los Angeles	15309	San Diego...	4798	Stanislaus..	6499
Calaveras....	8895	Marin........	6903	San Francis.	149482	Sutter........	5030
Colusa.......	6165	Mariposa....	4592	San Joaq'in.	21050	Tehama......	3387
ContraCosta	8461	Merced	2807	San LuisOb	4772	Trinity.......	3213
Del Norte...	2022	Mendocino..	7530	San Mateo...	8635	Tulare........	4521
El Dorado ...	10309	Mono........	430	Santa Barba	7784	Tuolumne...	8150
Fresno.......	4605	Monterey....	9881	Santa Clara.	26246	Yolo.........	9899
Humboldt..	2694	Napa........	5521	Santa Cruz.	4944	Yuba........ ..	10851
Inyo.........	1956	Nevada	19136	Shasta........	4173		
Kern	2925	Placer	11357	Sierra........	5939	Total	549808

MISSOURI.—Area, 67,380 square miles.

Adair....	11449	Davies......	14410	Macon	23230	Reynolds....	3756
Andrew......	15137	De Kalb......	9858	Madison......	5849	Ripley......	3175
Atchison....	8440	Dent	6357	Marles.......	5915	St. Charles .	21304
Audrain.....	12307	Douglas	3915	Marion......	28776	St. Clair.....	6742
Barry	10373	Dodge........	—	McDonald...	5226	St. Francois	9741
Barton	5087	Dunklin.....	5982	Mercer......	11557	Ste. Genev've	8384
Bates........	15960	Franklin....	30098	Miller........	6616	St. Louis ...	339774
Benton......	11322	Gasconade...	10093	Mississippi..	3713	Saline.........	21672
Bollinger....	8162	Gentry	11607	Moniteau...	11335	Schuyler.....	7987

Boone	20765	Greene	21549	Monroe	17149	Scotland	10670
Buchanan	26932	Grundy	10567	Montgom'ry	10405	Scott	7317
Butler	4298	Harrison	14635	Morgan	8434	Shannon	2339
Caldwell	11390	Henry	17401	New Madrid	6357	Shelby	10119
Callaway	19202	Hickory	6452	Newton	12821	Stoddard	8535
Camden	6108	Holt	11652	Nodaway	14751	Stone	3253
Cape Gir'd'n	17558	Howard	17233	Oregon	3287	Sullivan	11908
Carroll	17445	Howell	4218	Osage	10793	Taney	4407
Cass	19296	Iron	6278	Ozark	3363	Texas	9618
Carter	2100	Jackson	55041	Pemiscott	2059	Vernon	11246
Cedar	9474	Jasper	14929	Perry	9877	Warren	9673
Charlton	19135	Jefferson	15380	Pettis	18706	Washington	11719
Christian	6707	Johnson	24649	Phelps	10506	Wayne	6068
Clark	13667	Knox	10970	Pike	23076	Webster	10434
Clay	15564	Laclede	9380	Platte	17349	Worth	5004
Clinton	14063	Lafayette	22623	Polk	12445	Wright	5684
Cole	10292	Lawrence	13067	Pulaski	4714		
Cooper	20692	Lewis	15114	Putnam	11217	Total	1691693
Crawford	7982	Lincoln	14073	Ralls	10510		
Dade	8683	Linn	15900	Randolph	15908		
Dallas	8383	Livingston	10116	Ray	18700		

MINNESOTA.—Area, 95,274 square miles.

Atkin	18	Faribault	9390	McPhail	no ret.	St. Louis	4561
Anoka	3940	Fillmore	24887	Meeker	6100	Scott	11042
Beckel	308	Freeborn	10583	Mille Lacs	1109	Sherburne	2050
Beltrami	80	Goodhue	19214	Monong'lia	3161	Sibley	6725
Benton	1558	Grant	340	Morrison	1899	Stearns	14206
Big Stone	no ret.	Hennepin	31566	Mower	10448	Steele	8270
Blue Earth	17393	Houston	11661	Murray	259	Stevens	187
Brec'nr'ge	—	Isanti	350	Nicollet	8380	Todd	1818
Brown	6396	Itasca	78	Nobles	117	Toombs	—
Buchanan	—	Jackson	1825	Olmstead	19793	Traverse	no ret.
Carlton	286	Kanabac	93	Otter Tail	221	Wabashaw	15860
Carver	11587	Kandiyohi	1760	Pembina	64	Wadenah	6
Cass	184	Lac q.Parle	no ret.	Pierce	—	Wahnatah	—
Chisago	4358	Lake	135	Pine	648	Waseca	7854
Chippewa	no ret.	La Sœur	11607	Pipestone	no ret.	Washington	12490
Clay	92	Lincoln	3219	Polk	no ret.	Waterman	2426
Cottonw'd	534	Lyon	no ret.	Pope	2691	Wilkin	295
Crow Wing	200	Mankahta	no ret.	Ramsey	23086	Winona	22319
Dakota	13341	Manonim	no ret.	Redwood	1829	Wright	9457
Dodge	8598	Martin	3867	Reneville	no ret.		
Douglas	3500	McLeod	5643	Rice	16083	Total	424543

SOUTH CAROLINA.—Area, 29,385 square miles.

Abbeville	3129	Darlington	20000	Lexington	15000	Sumter	23000
Anderson	22000	Edgefield	40000	Marion	20000	Union	19000
Barnwell	30000	Fairfield	19888	Marlboro'gh	11819	Williamsb'g	15489
Beaufort	40000	Georgetown	16161	Newberry	17783	York	23000
Charleston	78000	Greenville	20000	Oconee	10503		
Chester	18000	Horry	10721	Orangeburg	25000	Total	705789
Chesterfield	10593	Kershaw	13000	Pickens	10269		
Clarendon	13000	Lancaster	12087	Richland	18000		
Calleton	40000	Laurens	23000	Spartanburg	25785		

NEW YORK.—Area, 47,000 square miles.

Albany	133109	Fulton	27056	Ontario	45220	Steuben	67996
Allegany	40764	Genesee	32209	Orange	81503	Suffolk	46960
Broome	44176	Greene	38403	Orleans	27822	Sullivan	34589
Cat'rug's	44924	Hamilton	2960	Oswego	78026	Tioga	30573
Cayuga	59513	Herkimer	39936	Otsego	48698	Tompkins	33168

106

Chaut'cua... 59126	Jefferson 64450	Putnam...... 12862	Ulster 83657
Chemung ... 35341	Kings.........420292	Queen........ 83847	Warren....... 22605
Chenango.... 40553	Lewis....... 28452	Rensselaer.. 99587	Washington 49342
Clinton 48622	Livingston.. 38321	Richmond... 33044	Wayne....... 47720
Columbia.... 47087	Madison..... 43595	Rockland ... 25163	W'tch'st'r....132288
Cortland.... 25220	Monroe......117462	St.Lawrence 84881	Wyoming.... 29176
Delaware.... 42982	Montg'mry. 34510	Saratoga..... 51513	Yates......... 19608
Dutchess.... 71887	New York..926341	Schen'ct'dy. 21348	
Erie 176930	Niagara...... 50894	Schoharie... 33289	Total.....4370846
Essex......... 29076	Oneida 110036	Schuyler.... 18219	
Franklin..... 30717	Onondaga...104404	Seneca........ 27844	

GEORGIA.—Area, 58,000 square miles.

Appling...... 5086	Dawson....... 3800	Jefferson..... 12190	Richmond .. 21200
Baker......... 4900	Decatur...... 11900	Johnson 2961	Schley....... 5129
Baldwin 10618	De Kalb...... 10014	Jones......... 9436	Schriven..... 8200
Banks........ 4700	Dooley 9790	Laurens...... 7834	Spaulding .. 10205
Bartow 16566	Dougherty.. 11517	Lee.......... 9567	Stewart 14205
Berrien 4518	Early......... 6998	Liberty...... 12229	Sumter...... 16559
Bibb 21255	Echolls 1400	Lincoln...... 5413	Talbot....... 11913
Brooke....... 8342	Effingham... 4700	Lowndes..... 8321	Taliaferro... 4793
Bryan........ 5252	Elbert........ 10400	Lumpkin.... 5161	Tatnall....... 4860
Bullock..... 5610	Emanuel.... 6134	Macon........ 11458	Taylor....... 7143
Burke 17100	Fannin....... 5429	Madison..... 5227	Telfair....... 2700
Butts......... 6941	Fayette...... 8221	Marion....... 8000	Terrell 9053
Calhoun...... 5503	Floyd......... 17233	McIntosh .. 4491	Thomas...... 10700
Camden...... 4611	Forsyth 7983	Merriw'ther 13756	Towns........ 2780
Campbell.... 9176	Franklin ... 7893	Muller 1700	Troup........ 17632
Carroll...... 11782	Fulton 33446	Milton........ 4281	Twiggs....... 8546
Cass .... —	Gilmer 6644	Mitchell,.... 6633	Union......... 5267
Catoosa.,..... 4409	Glasscock... 2400	Monroe...... 17213	Upson........ 9430
Chatham.... 41270	Glynn....... 5376	Montgomr'y 3586	Walker 9925
Chat'ho'chie 6059	Gordon...... 9268	Morgan...... 10696	Walton...... 11038
Chattooga... 6902	Greene 12454	Murray....... 6500	Warren...... 10545
Charlton.... 1897	Gwinnett.... 12431	Muscogee ... 16663	Ware 2286
Cherokee.... 10399	Habersham. 6322	Newton...... 14315	Washington 15842
Clarke...... 12911	Hall......... 9300	Oglethorpe.. 11782	Wayne....... 2177
Clay......... 5493	Hancock.... 11317	Paulding.... 7639	Webster...... 4677
Clayton...... 5477	Haralson ... 4004	Pickens...... 5317	White........ 3300
Clinch....... 3945	Harris....... 13284	Pierce........ 2778	Whitfield ... 10117
Cobb......... 13814	Hart 6783	Pike.......... 10905	Wilcox....... 2439
Coffee 3192	Heard........ 7866	Polk.......... 7822	Wilkes....... 11799
Columbus... 11800	Henry........ 10700	Pulaski....... 11940	Wilkinson... 9300
Colquitt...... 1654	Houston 20406	Putnam 10461	Worth........ 3787
Coweta...... 14700	Irwin......... 1837	Quitman 4150	
Crawford.... 7555	Jackson 10600	Rabun........ 3256	Total 1174832
Dade 3033	Jasper........ 10439	Randolph.... 9500	

TEXAS.—Area, 237,504 square miles.

Anderson ...	Duvall 866	Johnson......	Refugio...... 2325
Ange'ina ...	Eastland.....	Kornes........	Robertson...
Atacosta.....	Edwards....	Kaufman....	Rusk ....
Austin	Ellis..........	Kendall......	Sabine........
Bandera.....	El Passo	Kerr	St.Augustine
Bastrop......	Ensinal.......	Kinney	San Patricio 602
Bee	Erath.........	Lamar	San Saba....
Bell	Falls..........	Lampass.....	Shakleford..
Bexar	Faunin.......	LaSalle 69	Shelby........
Blanco.......	Fayette......	Lavaca....... 9168	Smith........
Bosque	Fort Bend ... 1541	Leon	Starr.........
Bowie	Freestone....	Liberty 4413	Tarrant.......
Brazoria 7528	Frio.............	Limestone...	Throckmort'n

County	Pop.	County	Pop.	County	Pop.	County	Pop.
Brazos		Galveston	15290	Live Oak	852	Titus	
Brown		Gillespie		Llano		Travis	
Buchanan		Gollad		McLennan		Trinity	
Burleson		Gonzales		McMullen	230	Tyler	5010
Burnett		Grayson		Madison		Upshur	
Caldwell		Grimes		Mariou		Uvalde	
Calhoun		Guadalupe		Mason		Van Zandt	
Camancye		Hamilton		Matagorda		Victoria	
Cameron		Hardeman		Maverick		Walker	
Cass		Hardin	1460	Medina		Washington	
Chambers	1503	Harris	1737	Milan		Webb	
Cherokee		Harrison		Montague		Wharton	
Clay		Hayes		Montgomery		Williamson	
Coleman		Henderson		Nacogdoches		Wise	
Collin		Hidalgo		Navarro		Wilson	
Colorado		Hill		Newton	2187	Wood	
Comal		Hood		Nueces	4193	Young	
Cooke		Hopkins		Orange	1255	Zapata	1488
Coryell		Houston		Palo Pinto		Zavola	
Dallas		Hunt		Panola			
Davis		Jack		Parker		**Total**	**795500**
Dawson		Jackson		Polk			
Denton		Jasper	4218	Presidio			
De Witt		Jefferson	1906	Red River			

ALABAMA.—Area, 50,722 square miles.

County	County	County	County	Pop.
Autauga	Covington	Madison	Shelby	
Baldwin	Dale	Marengo	St. Clair	
Barbour	Dallas	Marion	Sumter	
Bibb	De Kalb	Marshall	Tallapoosa	
Blount	Fayette	Macon	Talladega	
Butler	Franklin	Mobile	Tuscaloosa	
Calhoun	Greene	Montgomery	Walker	
Chambers	Henry	Monroe	Washington	
Cherokee	Jackson	Morgan	Wilcox	
Choctaw	Jefferson	Perry	Winston	
Clarke	Lawrence	Pickens		
Coffee	Lauderdale	Pike	**Total**	**996175**
Conecuh	Limestone	Randolph		
Coosa	Lowndes	Russell		

LOUISIANA.—Area, 41,255 square miles.

County	Pop.	County	Pop.	County	Pop.	County	Pop.
Ascension	11577	Condordia	9977	Morehouse	9431	St. J'hnBap.	8200
Assumpti'n	13247	De Soto	14962	Natchitoch	18265	St. Landry	24681
Avoyelles	12926	Feliciana E.	5623	Opelousas	——	St. Martin's	9370
Bat. Rou. E.	17820	FelicianaW.	10498	Orleans	191322	St. Mary's	13860
" " W.	7500	Franklin	5124	Onachita	11582	St. Tam'ny	5587
Bienville	10644	Grant	9590	Plaqu'min's	10557	Tangipao	7928
Bossier	12675	Iberia	9042	Point Coup.	12981	Tensas	12419
Caddo	21719	Iberville	12347	Rapides	18015	Terrebonne	12451
Calcasieu	6733	Jackson	7646	Richland	5110	Union	11684
Caldwell	4824	Jefferson	17767	Sabine	6457	Vermillion	4528
Cameron	1591	Lafayette	10388	St. Bernard	3553	Washingt'n	3330
Carroll	10110	Lafourche	14723	St. Charles	4868	Winn	4959
Catahoula	8474	Livingston	4029	St. Heleua	5423		
Claiborne	20240	Madison	16000	St. James	10153	**Total**	**734420**

INDIANA.—Area, 33,809 square miles.

County	Pop.	County	Pop.	County	Pop.	County	Pop.
Adams	11342	Fulton	12717	Marion	65296	Shelby	21889
Allen	46416	Gibson	17353	Marshall	20377	Spencer	18001
Bartholm'w	22211	Grant	18499	Martin	11089	Starke	3890
Benton	5642	Greene	19192	Miami	21055	Steuben	12854

Blackford...	6266	Hamilton....	20894	Montg'mry.	23764	Sullivan......	18351
Boone.........	22593	Hancock.....	15112	Monroe......	14193	Switzerland	12131
Brown........	8581	Harrison.....	20005	Morgan......	17474	Tipton........	11953
Carroll........	16154	Hendricks...	20402	Newton......	5826	Tippecanoe.	34703
Cass...........	24191	Henry........	23036	Noble.........	20391	Union........	6343
Clinton	17339	Howard......	15850	Ohio...........	5837	Vanderberg.	33146
Clay	19086	Huntington	19033	Orange.......	13491	Vernillion..	10893
Clarke........	24116	Jackson.....	19413	Owen.........	16216	Vigo.........	34554
Crawford....	9852	Jasper........	6353	Parke.........	18195	Wabash......	21313
Daviess	16742	Jay...........	15000	Perry.........	14759	Warren	10207
Dearborn....	24118	Jefferson.....	29737	Pike...........	10844	Warwick...	14568
Decatur......	18888	Jennings.....	16212	Porter........	13038	Washington	17497
De Kalb....	16176	Johnson......	18404	Posey.........	19185	Wayne.......	31865
Delaware....	25284	Knox........	21575	Pulaski	7822	Wells........	13573
Dubois.......	12596	Kosciusko...	23929	Putnam......	21508	White	10772
Elkhart	25993	Lagrange....	11146	Randolph ...	22878	Whitley......	14501
Fayette......	10494	Lake........	12352	Ripley	20980		
Floyd.........	23109	Laporte......	27061	Rush..........	17621	Total......	1655675
Fountain....	16421	Lawrence ...	14497	St. Joseph...	25287		
Franklin.....	19543	Madison......	22772	Scott.........	7873		

KENTUCKY.— Area, 37,680 square miles

Adair........	11065	Estill	9198	Laurel........	6015	Pendleton...	14030
Allen.........	10296	Fayette	26656	Lawrence....	8497	Perry.........	4274
Anderson ...	5419	Fleming......	12918	Lee...........	3055	Pike...........	9562
Ballard	12576	Floyd.........	7877	Letcher......	4608	Powell.......	2599
Barren........	17780	Franklin.....	15300	Lewis.........	9000	Pulaski	17669
Bath	9566	Fulton........	6161	Lincoln	10947	Robertson...	5514
Boone........	10696	Gallatin......	5074	Livingston..	8200	Rockcastle..	6535
Bourbon....,	14863	Garrard	10376	Logan........	20429	Rowan	2991
Boyd........	8507	Grant........	9529	Lyon.........	6233	Russell.......	5809
Boyle.........	9515	Graves........	19399	McCracken .	13988	Scott.........	11607
Bracken	11409	Grayson.....	11580	McLean......	7613	Shelby.......	15733
Breathitt.....	5672	Greene	9379	Madison.....	19543	Simpson......	9573
Breckenr'ge	13441	Greenup.....	11463	Magoffin	4684	Spencer	5956
Bullitt........	7781	Hancock.....	6581	Marion.......	12840	Taylor.......	8267
Butler	9404	Hardin.......	15705	Marshall....	9457	Todd	12612
Caldwell	10826	Harlan.......	4415	Mason	18127	Trigg	13686
Callaway...	8262	Harrison.....	12877	Meade	9486	Trimble......	5577
Campbell....	27406	Hart...........	13687	Menifee......	1986	Union........	13640
Carroll........	6189	Henderson..	18457	Mercer.......	13145	Warren	20761
Carter	7509	Henry........	11066	Metcalfe	7934	Washington	12164
Casey	8884	Hickman....	8453	Monroe......	9231	Wayne.......	10602
Christian ...	20802	Hopkins	13827	Montg'mery	7557	Webster......	10937
Clarke	10882	Jackson	4547	Morgan	5975	Whitley......	8278
Clay..........	7100	Jefferson.....	118956	Muhlenberg	12638	Wolfe.........	3603
Clinton	6381	Jessamine...	8638	Nelson	14804	Woodford ...	8240
Crittenden..	9382	Johnson.....	7498	Nicholas	9129		
Cumberland	7690	Josh Bell....	3731	Ohio...........	22638	Total......	1320407
Daviess	20714	Kenton	36096	Oldham......	9027		
Edmonson...	4459	Knox.........	8276	Owen	14309		
Elliott........	4433	Larue........	8235	Owsley.......	3889		

FLORIDA.—Area, 59,268 square miles.

Alachner....	17348	Gadsden......	9794	Madison......	11023	Sumter	2952
Baker.........	1325	Hamilton ...	5739	Manatee	1899	Taylor........	1452
Bradford.....	3671	Hernando...	2939	Marion......	—	Volusa......	1723
Brevard.....	1216	Hillsbor'ugh	3216	Monroe	—	Wakulla.....	2506
Calhoun.....	998	Holmes.......	—	Nassau......	4259	Walton	—
Clay.........	2099	Jackson	9527	Orange.......	2194	Washington	2302
Columbia ...	7325	Jefferson.....	13410	Polk........ ..	3196		
Dade..........	—	Lafayette ...	1873	Putnam......	3821	Total	189995
Duvall	11932	Leon.........	15337	Santa Rosa..	—		
Escambia....	—	Levy.........	2018	St. John's...	2618		
Franklin.....	2256	Liberty	1052	Suwanee.....	3591		

KANSAS.—Area, 78,418 squars miles.

Allen	7023	Greenwood	3484	Neosho	9610	Woodson	3827
Anderson	5220	Godfrey	—	Norton	—	Wyandott	10019
Atchison	15507	Gove	—	Osage	7648		
Barbour	—	Harper	—	Osborne	33	Total	379497
Barton	2	Hodgeman	—	Otoe	—		
Bourbon	15076	Howard	2794	Ottawa	2127	**NEBRASKA.**	
Breckinr'ge	—	Jackson	6053	Pawnee	179	Total	116888
Brown	6824	Jefferson	12526	Phillips	—		
Butler	3035	Jewell	207	Pott'wat'mie	7848	**NEVADA.**	
Chase	1980	Johnson	13684	Pratt	—	Churchill	196
Cherokee	11047	Kiowa	—	Reno	—	Douglass	1215
Clarke	—	Labette	9496	Republic	1281	Esmeralda	1553
Clay	2942	Leavenw'th	32444	Rice	5	Elko	3447
Cloud	2323	Lincoln	516	Riley	5105	Humboldt	1916
Coffee	6210	Linn	12174	Rooks	—	Lander	2815
Comanche	—	Lykins	—	Rush	1	Lincoln	2185
Cowley	1175	Lyon	8024	Russell	156	Lyon	1837
Crawford	8160	Madison	—	Saline	4246	Nye	1087
Davis	3993	Marion	768	Sedgwick	1095	Ormsby	3668
Dickinson	3043	Marshall	6904	Shawnee	15121	Pahute	765
Doniphan	13969	McGhee	—	Smith	66	Roop	133
Dorn	—	McPherson	738	Stafford	—	Storey	11359
Douglas	20604	Miami	11725	Sumner	—	Washoe	3091
Ellis	1336	Mitchell	488	Trego	166	White Pine	7189
Ellsworth	1185	Montgom'ry	7564	Wabaunsee	3362		
Ford	—	Morris	2225	Wallace	538	Total	42456
Franklin	10085	Nemaha	7339	Washington	4081		
Graham	—	Ness	2	Wilson	6694		

TERRITORIES.

Arizona.
Mohave...... 179
Pima.......... 5716
Yavapai...... 2142
Yuma......... 1621

Total 9658

Colorado.
Arapahoe ... 6829
Bent 592
Boulder...... 1939
Clear Creek 1596
Conejos...... 2479
Costello...... 1779
Douglas 1388
El Paso...... 987
Fremont 1064
Gilpin 5490
Greenwood . 510
Huerfano ... 2250
Jefferson.... 2390
Lake.......... 522
Larimer...... 838
Las Animas 4276
Park. 447
Puebla....... 2265
Sagnache.... 304
Summit 258
Weld'......... 1478

Total 39681

Dakota.
Bon Homme 608
Brookings... 163
Buffalo....... 246
Charles Mix 152
Clay.......... 2621
Denel 37
Hutchison... 37
Jayne.........
Lincoln...... 712
Minnehaha. 355
Pembina 1213
Todd.......... 337
Union......... 3507
Yankton..... 2097
Unorganized 2091

Total 14181

Idaho.
Ada. 2675
Alturas....... 685
Boise 3833
Idaho......... 849
Lemhi 988
Nez Perces.. 1495
Oneida 1922
Owyhee...... 1715
Shoshone.... 722

Total 14882

Montana.
Beaver Head 722
Big Horn.... 38
Chouteau.... 517
Dawson....... 177
Deer Lodge. 4364
Gallatin,..... 1579
Jefferson 1531
Lewis & Clark 5041
Madison...... 2684
Meagher...... 1387
Missoula..... 2554

Total 20594

New Mexico.
Arizona......
Bernalillo... 7569
Colfax........ 1993
Dona Anna.. 5863
Mora......... 8056
Rio Arriba.. 7641
Santa Anna 2124
Santa Fe 8930
San Miguel. 16021
Socorro 6410
Taos 9707
Valentia 8925

Total 86122

Utah.
Total 70000

Washington.
Chehalis..... 401
Clallam...... 408
Clark.......... 3081
Cowlitz 730
Island......... 626
Jefferson..... 1268
King.......... 2120
Kitsap ,...... 866
Klikitat...... 329
Lewis 888
Mason 289
Pacific........ 738
Pierce......... 1409
Skamania.... 133
Swawamish 575
Stevens...... 734
Thurston.... 2246
Waukiakum 270
Walla Walla 5300
Whatcum.... 534
Yakima...... 432
Disputed Isls 524

23901
Indians...... 24

Total 23925

Wyoming.
Albany....... 2021
Carbon....... 1368
Laramie...... 2957
Sweetwater 1916
Uintah....... 856

Total 9118

Di. Columbia.
Georgetown 11385
Washingt'n 109204
Bal. of Dis't 11117

Total 131706

INDEX.

A

Air, qualities of, 31; capacity for moisture, 34; vapors in spaces filled with, 34; composition of, 34.
Amazon river, 43
Amending and appealing, 19
Answers to letters, 3
Application by actual inventor, 17; fee on, 7; signing and witnessing, 12; not refused because known in another country, 18
Assignee of invention, 7
Assignment, recorded, 7; requirements in regard to, 24
Asteroids, 43
Atmosphere, weight of, 38; height of, 38
Atlantic ocean, depth of, 43
Australian patents, 64
Austrian patents, 70
Axle grease, 31

B

Bath metal, composition, 43
Bavarian patents, 68
Belgium, patents in, 60
Bell metal, composition, 50
Benzole, 43
Book, object of, 3
Brass, coloring, 41
Bronzing wood, 44
Burns, remedy for, 51

C

Cannon ball, velocity of, 31
Carbon, 41
Caveat, 9; fee and receiving precedence, 9
Cement, jewellers, 35; for steam pipes, 42
Ceylon, patents in, 70
Circle, area of, 12
Cities of the U. S., population of, 58
Clouds, height of, 33; the cirrus composed of flakes of snow, 33
Coal districts, products of, 50
Cold, artificial, 41
Colonies of England, patents in, 69
Compass, 40
Copyrights, cost, how obtained, 18
Coral, artificial, 38

Cords, strength of, 38
Correspondents, 10
Cuban patents, 66
Census, by counties, of every state, 99

D

Day and night, length of, 47
Denmark, patents in, 69
Dew, production of, 33; effect on the thermometer, 34
Diamond, refraction of, 36
Disclaimers, nature of, 24
Drawings of inventions, 27-28; rendered permanent, 48

E

Earth, the weight of, 42; density of, 50
East Indian patents, 69
Electricity, speed of, 12
Elements, number of, 40
Ellipsis, area of, 16
English patents, 63
Extensions, given when, cost etc., 22

F

Fac-similes of signatures, etc., 49
Fuel, effect of different kinds, 45; table of power, 47
French patents, 63

G

Gases, expansion of, 34; absorption of heat, 50
German principalities, patents in, 69; German silver, 51
Glue, 41; waterproof, 42
Gold lining, 39
Gravity, force of, 36; to find specific 41
Grease, removing, 35
Greece patents in, 69
Gun Cotton, 49
Gunpowder, force of, 86; composition of, 33
Gutta-percha, to dissolve, 14

H

Hearing, 25
Heat, transmission of by various substances, 29; non-conductors, 36; penetration of sun's heat, 38; heat of the human body, 49

INDEX.

www.ingramcontent.com/pod-product-compliance
Lightning Source LLC
Chambersburg PA
CBHW022047210326
41519CB00055B/1103